# THE CONNECTIVITY CRISIS

## CRISIS

## Half the World Left Behind

# THE CONNECTIVITY CRISIS

## Half the World Left Behind

Essential Strategies to Connect One-Third of the Global Population Currently Offline & Meaningfully Connect a Further 2 billion of the So-Called Connected.

Expert Insights for Global Institutions and Developing Nations Committed to Achieving Universal Digital Inclusion.

## H. Sama Nwana

Visiting Professor
Department of Electronic & Electrical Engineering
University of Strathclyde
Glasgow, Scotland, UK.

First Published in October 2025 by Strathclyde Academic Media.

All hyperlinks in this book were active on 15th August 2025.

The author and publisher of this book have made best efforts to contact copyright holders to seek permissions where appropriate. In the event of any omissions or errors, please get in touch via the above email address, and corrections will be made for future editions.

*The full text of this book has been peer-reviewed to ensure high quality academic standards.*

Edited by Louise Crockett.

**THE CONNECTIVITY CRISIS: Half the World Left Behind**

*H Sama Nwana*

Paperback ISBN:    978-1-7395886-4-9

Hardback ISBN:    978-1-7395886-5-6

# Connecting the 2.6 billion *not* Online & *meaningfully* connecting 2 billion more

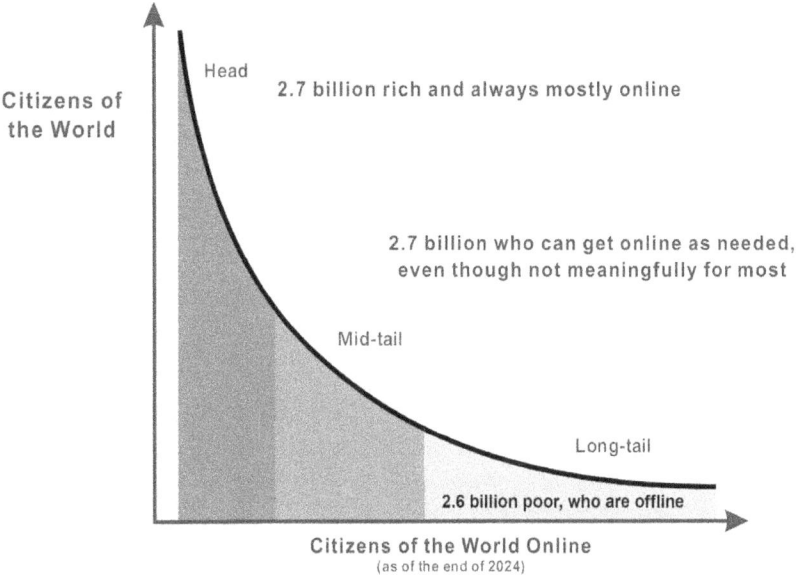

Citizens of the World

Head

2.7 billion rich and always mostly online

2.7 billion who can get online as needed, even though not meaningfully for most

Mid-tail

Long-tail

2.6 billion poor, who are offline

Citizens of the World Online
(as of the end of 2024)

## *Unsolicited* Insights, Advice and Guidance to Developing Countries, the Broadband Commission, the ITU, the World Bank Group, UN Agencies and more.

"Don't confuse having less with *being less*, having more with *being more*, or what you have with *who you are*."

Noah benShea[1]

[1] https://www.noahbenshea.com/about/

# Dedication

In memory of my late twin, *Protus Sama Nwana (1964 – 1994)*; not a day passes without you being in my thoughts.


In memory of my late mum *Odilia Mantan Nwana (1942 – 2021)*. I urge her descendants to remember and adhere to the fundamental principles of her life's philosophy and works: <u>treat others with kindness; strive to be your best self; do what is right; and do the right things both excellently and stubbornly</u>[2]. I am trying Mama.

---

[2] Like my late mother, there are three key elements to bequeath to future generations: *connection, integrity,* and *values.* It is my aspiration that all my mother's descendants will cherish their bond [connection] with her, cultivate her integrity, and strive to embody the values she practiced.

# Acknowledgments

I am indebted to so many colleagues and ex-colleagues who have influenced my thoughts and views on the connectivity crisis over more than a decade across the Ofcom UK, the Dynamic Spectrum Alliance, Cenerva Ltd, the University of Strathclyde and more. Their knowledge and works have guided some of how I chose to articulate many concepts in this book.

Particularly, I acknowledge Paul Garnett (formerly of Microsoft), who not only founded the Dynamic Spectrum Alliance[3] which strives to promote radio spectrum policies and regulations that bridge the digital divide, but who continues to be a devout advocate and worker towards addressing the digital connectivity deficit issues. I also particularly acknowledge Mr Eric J. Wilson who co-inspired and co-developed some insights in this book, particularly the 'Civilisation in a Box' proposition of Chapter 4.

I am exceedingly grateful for immensely helpful, and sometimes incredibly detailed, peer-review comments on the manuscript that evolved into this book from Dr Charley Lewis, Professor William Webb, Dr Shruthi K. Anantha Kumar, and Professor Alison Gillwald.

I am incredibly indebted to Dr Shruthi Koratagere Anantha Kumar who principally authored an opinion piece[4] with me (as coauthor) which I have largely republished as Section 7.2 of this book. I have done so with her full permission.

I am so indebted to Ms Catherine Dawson who designed the book cover and came up with the book title too. I thank Mr Gana Nwana who also reviewed the entire manuscript for errors.

---

[3] https://www.dynamicspectrumalliance.org/
[4] https://cenerva.com/wp-content/uploads/2025/02/BharatNet-report.pdf

Finally, I extend my gratitude to Strathclyde Academic Media for including this book in its publication portfolio. I particularly wish to acknowledge Dr Louise Crockett, who edited the book, and Professor Bob Stewart who encouraged me to publish with Strathclyde Academic Media. Both are full time academics with the University of Strathclyde, Glasgow, Scotland.

All remaining errors are therefore of course my own.

# Preface

At the outset, I acknowledge that this book does *not* adhere to the conventional impersonal and formal academic style. Instead, it is intentionally crafted in a more personal and conversational manner, as noted by peer reviewers. This approach aims to better engage the intended readership, particularly those with limited time (i.e., very busy and therefore time poor), by allowing them to access key messages efficiently and at their convenience. This Preface, Chapter 1 and the concluding chapter, "The Book Summarised," provide an overview of the book's central themes, while the remaining chapters offer further detail.

Universal online digital inclusion is this *nirvana* that the United Nations (UN) and all its 194-odd Member States are [supposedly] striving for. This is because being *connected online* has become an essential utility for daily living such as working, being entertained, accessing information like news and health information, accessing basic services provided by the Government and others, staying in touch with friends and family, increasing SME[5] productivity, self-actualisation (like me authoring this book), and engaging in other forms of self-empowerment.

The United Nations recognises Internet access as a fundamental human right, notably through the Human Rights Council's resolutions. The key resolution, *"The promotion, protection and enjoyment of human rights on the Internet"* was first adopted in 2012 and updated significantly in 2021[6]. Amongst several key tenets of the resolution, it emphasises bridging the digital divide, calling on Governments to adopt inclusive policies that ensure universal, affordable Internet access. Whilst this resolution is not legally binding, it carries -or should carry] significant moral and political weight.

To illustrate how important - or more accurately, how *meaningful* – Internet access (inclusive of digital inclusion and online connectivity) is, let us consider some online usage data from the UK. The UK Telecommunications, Media and

---

[5] Small and Medium Enterprises
[6] https://www.article19.org/resources/un-human-rights-council-adopts-resolution-on-human-rights-on-the-internet/

Technology (TMT) regulator [Ofcom] reported in 2023 that '*we're spending an extra two days each year online*'[7]. Ofcom reported that:

> "the average amount of daily time spent online has seen a modest increase of eight minutes, to *3 hours and 41 minutes* a day in May 2023. This means the average online adult now spends around *56 days each year online* – two more days than in 2022". Young online adults aged 18–24-spend the most time online, at *4 hours 36 minutes each day in May 2023*".
>
> *(the author's emphasis)* – Source:[8]

The average UK adult spent 56 days online in 2023! From the above citation, young adults spend – by my math – almost 70 days a year online in the UK, or circa 20% of their lives[9]. I do *not* know about you, but I find this both incredible and worrying. I do not intend to undermine my connectivity imperative argument here with this following digression. However, it is already evident the many unhealthy consequences that have already accrued (and is accruing daily) including children *not* having enough sleep, mental health challenges, children encountering sexual risks online, emotional and psychological harms and more[10]. This and more are the subject of a book I wrote in three years ago (Nwana, 2022). End of digression.

The core question of this book is the following. Why should UK citizens be so digitally-connected and included, and be able to use online services so much – yet billions of others around the world are ***not*** equally *meaningfully* included – or not digitally-connected online at all? Combined (i.e. the unconnected and under connected), half the world is being left behind! I return to this issue of *meaningful* connectivity in Chapter 1 and other chapters. These core broad issues are really at the genesis of this book.

---

[7] https://www.ofcom.org.uk/media-use-and-attitudes/online-habits/top-trends-from-our-latest-look-at-peoples-online-lives

[8] *Ibid.*

[9] One reviewer cleverly questioned: "is this 'active' browsing or 'passive' waiting for the next WhatsApp ping!"?

[10] https://learning.nspcc.org.uk/research-resources/2023/online-risks-to-children-evidence-review

However, the unequivocal top-level vision of this book is to digitally-connect online *everybody*[11] in the world, and this is broadly termed the *digital inclusion challenge*.

The reality, according to the International Telecommunications Union (ITU)[12] as of the end of 2024[13], is that there are 2.6 billion people who are still offline. According to the ITU's 2022 Facts & Figures[14], the distribution across regions of the 2.7 billion then-unconnected to the Internet and the digital realm was as follows: the majority of the unconnected reside in Africa (60%), followed by the Asia-Pacific (36%), the Arab States (39%), the Americas (17%), CIS (16%), and Europe (11%). These ratio percentages are still broadly the same as of early 2025. The slow rate shows how hard this connectivity nut is to crack.

So, in this book, *The Connectivity Crisis - Half the World Left Behind*, I make many recommendations and propose 'out of the box', fit-for-purpose, implementable approaches and solutions that UN member states could adopt. It is different thinking, a unique set of insights and an approach borne of the realities that the 2.6 billion unconnected 'live' today.

I hope - in particular - that Presidents/Prime Ministers and ICT Ministers of LICs, LMICs, SIDS, LLDCs and LDCs generally across Africa, Asia-Pacific, Arab States, Latin America and CIS leaf through some of these pages. I also hope that organisations like the ITU Broadband Commission, the ITU itself, the United Nations, UNESCO, the World Bank in addition to *all* LDC member states - not *only* take such insights, guidance and proposed solutions in this book seriously - but positively 'adopt' and own them.

Otherwise, their well-meaning intentions and visions to online-include all by 2030 – as per UN Agenda 2030[15] - will *never* be realised.

*H Sama Nwana*
*August 2025*

---

[11] Of course, children of a 'regulation-decided' age and below would be excluded here.
[12] https://www.itu.int/en/mediacentre/Pages/PR-2024-11-27-facts-and-figures.aspx
[13] https://www.itu.int/en/ITU-D/Statistics/pages/stat/default.aspx
[14] https://www.itu.int/itu-d/reports/statistics/facts-figures-2022/index/
[15] https://www.sightsavers.org/policy-and-advocacy/global-goals/

# Contents

# List of Abbreviations & Definitions

| | |
|---|---|
| 1G – 5G | Generations of mobile technology |
| 2G | Second Generation |
| 3G | Third Generation |
| 3GPP | Third Generation Partnership Project |
| 4G | Fourth Generation |
| 4G/LTE | Fourth Generation Long-Term Evolution. |
| 5G | Fifth Generation |
| 6G | Sixth Generation |
| AI | Artificial Intelligence |
| AP | Asian Pacific Telecommunity |
| ARPU | Average Revenue Per User |
| ATU | African Telecommunications Union |
| CARICOM | Caribbean Community |
| CBRS | Citizens Broadband Radio Service |
| CEO | Chief Executive Officer |
| Connected | 'The connected' or 'connected population' in this book refers to people who use Internet at least. 'The unconnected' refers to those who do not use Internet. |
| Coverage | 'Population coverage' is the share of the population that lives under signals provided by a mobile network that is strong enough to use telecoms services (voice, SMS, data). 2G, 3G or 4G coverage levels provided by networks are independent from each other. |
| Coverage Gap | Populations who do not live within the footprint of a mobile [broadband/voice] network. |
| CSR | Corporate Social Responsibility |
| D2C | Direct to Cell |
| D2D | Direct to Device |
| DOCSIS | Data Over Cable Service Interface Specification |
| DRC | Democratic Republic of Congo |
| DSA | Dynamic Spectrum Access |
| DSL | Digital Subscriber Line |
| DSIT | Department of Science, Innovation and Technology |

| | |
|---|---|
| DTT | Digital Terrestrial Television |
| EDC | Emerging and Developing Country |
| EU | European Union |
| FAAAM | Facebook (Meta), Amazon, Alphabet, Apple and Microsoft |
| FAANG | Facebook (Meta), Amazon, Apple, Netflix and Google |
| FANGAM | Facebook(Meta), Amazon, Netflix, Google, Apple and Microsoft |
| FDD | Frequency Division Duplex |
| FWA | Fixed Wireless Access |
| FSOC | Free-Space Optical Communications |
| GAMMA | Google, Apple, Microsoft, Meta, and Amazon |
| GEO | Geostationary Equatorial Orbit (Satellite) |
| G2P | Government-to-Person [payments] |
| GDP | Gross Domestic Product |
| GNI | Gross National Income |
| GSM | Global System for Mobile communications |
| GSMA | GSM Association |
| HAPs | High Altitude Platforms |
| HIC | High Income Countries |
| HIBs | High Altitude Platform Stations as IMT Base Stations |
| ICT | Information and Communications Technologies |
| IFC | International Finance Corporation |
| IMT | Internation Mobile Telecommunications |
| IoT | Internet of Things |
| IP | Internet Protocol |
| ISP | Internet Service Provider |
| ISRO | Indian Space Research Organisation |
| ITU | International Telecommunications Union |
| IXP | Internet eXchange Point |
| KPIs | Key Performance Indicators |
| LEO | Low Earth Orbiting (Satellite) |
| LDC | Least Developed Countries |
| LIC | Low-Income Countries |
| LMIC | Low-and Middle-Income Countries (LMIC) |
| LLDC | LandLocked Developing Countries (A country classified as landlocked and developing by the UN.) |
| LEO | Low Earth Orbit (Satellite) |

| | |
|---|---|
| LTE | Long Term Evolution |
| M2M | Machine to Machine |
| Mobile broadband | 3G, 4G or 5G technologies. |
| MDB | Multilateral Development Bank |
| M&E | Measurement & Evaluation |
| MEO | Medium Earth Orbiting (Satellite) |
| MFS | Mobile Financial Services |
| MHz | Megahertz |
| MNO | Mobile Network Operator |
| MSMEs | Micro-, Small- and Medium-sized Enterprises |
| MVNO | Mobile Virtual Network Operator |
| PAYG | Pay-As-You-Go |
| P2G | Person-to-Government [payments] |
| PLC | Power Line Communications |
| PV | PhotoVoltaic |
| QED | Quod Erat Demonstrandum – literally translated as "that which was to be demonstrated", or - more clearly as - "Just what we set out to prove" |
| NGO | Non-Governmental Organisation |
| NICTA | National Information & Communications Technology Authority |
| NRA | National Regulatory Authority |
| Ofcom | Office of Communications (UK) |
| OTT | Over the Top |
| QoS | Quality of Service |
| RPU | Remote Power Unit |
| SDG | Sustainable Development Goal |
| SDR | Special Drawing Rights |
| SIDS | Small Island Developing States (A country classified as a small island and developing state by the UN.) |
| SIM | Subscriber Identification Module |
| SMART | Specific, Measurable, Actionable, Realistic and Time-Bound |
| SMART-ER | Specific, Measurable, Actionable, Realistic, Time-Bound, Ethical and Reviewable |
| SME | Small and Medium Enterprise |

| | |
|---|---|
| SMS | Short Message Service |
| SRN | Shared Rural Network |
| SSA | Sub-Saharan Africa |
| TCO | Total Cost of Ownership |
| TMT | Telecoms, Media and Technology |
| TVWS | TV White Space |
| UAS | Universal Access Service |
| UMIC | Upper-Middle Income Countries |
| UNDP | United Nations Development Programme |
| UNECA | United Nations Economic Commission for Africa |
| UNESCO | United Nations Educational, Scientific and Cultural Organisation |
| UNICEF | United Nations International Children's Emergency Fund |
| Usage Gap | Populations who live within the footprint of a mobile [voice/broadband] network but do not use mobile internet. |
| USD | United States Dollar |
| USF | Universal Service Fund |
| USO | Universal Service Obligation |
| VOIP | Voice over Internet Protocol |
| VoLTE | Voice over LTE (Long Term Evolution) |
| VSAT | Very Small Aperture Terminal |
| WACC | Weighted Average Cost of Capital |
| WDC | Wholesale Dedicated Capacity |
| Wi-Fi | Wireless Fidelity |
| WISP | Wireless Internet Service Provider |

# Part I – The *Unrealisable* Digital Inclusion and Meaningful Connectivity Challenge for 2.6+ Billion

This **Part I** of the book consists of three chapters.

They address why the universal digital inclusion challenge is *nowhere* close to being realised for 2.6+ billion people. I assert that it is 2.6 billion+ (i.e., more than 2.6 billion) because I estimate there are more than 2 billion more who are considered to be connected to the Internet, but who *hardly* experience any 'meaningful connectivity'.

So, this part of the book introduces the online digital inclusion challenge as well as the notion of meaningful connectivity. It provides some real-world realities of the nature of both *the universal digital [online] inclusion and meaningful connectivity challenge,* some new insights [drawing from the Pareto principle and more]. It also proposes a set of characteristics required of 'out-of-the-box' approaches and/or solutions to the sustainable resolution of these intractable digital inclusion and meaningful connectivity challenges.

# 1 Introduction

The long form title of this chapter is '*Introduction: the digital inclusion challenge, meaningful connectivity and the Pareto principle*'.

It is *unarguable* today in early 2025 that – not millions – but billions of people worldwide remain offline. The current reality, according to both the ITU Broadband Commission[16] and the International Telecommunications Union (ITU)[17] itself, is that there is a *stubborn*[18] 2.6 billion people, as of the end of both 2023 and 2024[19], who are still offline. The danger here is that this stubborn number is flatlining or flattening off. Of the 8+ billion people on earth, this makes circa 32.5% of humanity being offline as of early 2025.

If you add to this number those who are not *meaningfully* connected today, I believe this percentage number is closer to 50% of humanity - if not more – who are *not* connected at all or not meaningfully connected.

Commenting on the ITU Facts and Figures 2024 report[20], ITU Secretary-General (SG), Doreen Bogdan-Martin, noted:

> "Facts and Figures 2024 is a *tale of two digital realities between high-income and low-income countries*. Stark gaps in critical connectivity indicators are cutting off the most vulnerable people from online access to information, education and employment opportunities. This report is a

---

[16]https://www.broadbandcommission.org/publication/state-of-broadband-2023/
"The Broadband Commission for Sustainable Development was originally launched in May 2010 by the International Telecommunication Union (ITU) and the United Nations Educational, Scientific and Cultural Organisation (UNESCO) as the Broadband Commission for Digital Development. It is comprised of top industry leaders, government leaders, international agencies and development organisations. Commission members work together to devise strategies that advocate for higher priority to be given to the development of broadband infrastructure and services, to ensure that the benefits of these technologies are realised in all countries, by all people" source: BroadbandCommission.org.
[17] https://www.itu.int/en/mediacentre/Pages/PR-2024-11-27-facts-and-figures.aspx
[18] My emphasis, my word!
[19] https://www.itu.int/en/ITU-D/Statistics/pages/stat/default.aspx
[20] *Ibid.*

reminder that true progress in our interconnected world isn't just about how fast we move forward, but about making sure everyone moves forward together."

I agree with the ITU SG that it is indeed a tale of two realities, between High Income Countries (HICs) vs. Low Middle-Income Countries/Low Income Countries (LMICs/LICs). Connecting the tail 2.6 billion people is more about a humongous 'load lift' online of the latter, and in LMICs/LICs mostly. It is also a tale of urban vs. rural divide: the ITU finds that of the 2.6 billion people offline in 2024, 1.8 billion people live in rural areas[21].

Regarding the 32.5% 'load lift', there are both positive and negative aspects to consider. The positive aspect is that 67.5% of the global population is already digitally connected and online, although this does not necessarily mean meaningful connectivity for many hundreds of millions, if not billions – further details on this estimate are discussed in Chapter 6. The negative aspect is not just the 32.5% that still need to be connected, but the significantly more *exponential* task of getting this third of the world's 8+ billion population online.

I provide the insight here that this exponentiality challenge *is well-predicted by the well-known Pareto principle*, also known as the 80/20 rule. This principle refers to the typically observed reality that with 20% of the effort you will reach 80% of your goal. However, to achieve the last 20% you need 80% of the effort. I note that the Pareto principle is *not* scientific QED[22], but a rule of thumb that seems to always hold. I write this to address the rightfully sceptical reader at this juncture.

Clearly, we have *not* even arrived 80% digital online inclusion goal yet in 2025, leaving only 20% *to be* digitally included. We are currently at 67.5% as noted earlier.

Simply put, Pareto suggests that connecting the remaining 2.6 billion offline people will require over 80% of the total effort. So, we have not even engaged

---

[21]  https://www.itu.int/en/ITU-D/Statistics/pages/stat/default.aspx
[22]  Q.E.D. stands for the Latin phrase *quod erat demonstrandum,* literally translated as "that which was to be demonstrated", or - more clearly as -"Just what we set out to prove".

80%+ of the effort *yet* to connect the long tail of 2.6 billion people. We must then look for 'smarter' and 'out-of-the-box' 'levers' or 'pulley'-type solutions[23] and/or other innovative solutions to *lessen* the 'load lift' of such humongous efforts still ahead of us. This is one of the core messages of this book.

This reality is what – I hope - makes the contribution of this book unique. We need 'out-of-the-box' thinking to *lessen* such a 'load lift' of the connection of this stubborn 2.6 billion, *in addition* to meaningfully connecting hundreds of millions, if not billions, more.

The challenge of achieving digital inclusion for all is significant. Despite efforts, only about 67.5% of the world was digitally included by the end of 2023 and 2024, according to the ITU Broadband Commission and the International Telecommunications Union (ITU[24]). In this chapter, I emphasize that most of the work to connect the remaining 2.6 billion people still lies ahead.

## 1.1   The Universal Digital Inclusion Challenge

For the past twenty to thirty years, the ITU, the Broadband Commission[25], the Internet Society[26], multinational agencies (e.g., the World Bank[27] and the IFC[28]) and major private multinationals (e.g., Google, Meta/Facebook, Amazon, Microsoft, etc.) have been striving to bridge the digital divide, to ensure access to broadband communications and the Internet as 'basic human rights' and to digitally-connect everybody in the world broadly. The ITU has been leading on its 'committed to connecting the world' motto principally through its World Telecommunication Development Conference (WTDC) series, which commenced 30+ years ago with the March 1994 WTDC-94 conference in Buenos Aires, Argentina.

The Broadband Commission strives to put universal broadband connectivity [for all humans on earth] at the forefront of global policy discussions through developing practical and sustainable policy recommendations to accelerate

---

[23] The idea of 'levers' and 'pulleys' solutions is explained in Chapter 3.
[24] https://www.itu.int
[25] https://www.broadbandcommission.org/
[26] https://www.internetsociety.org/
[27] https://www.worldbank.org/
[28] https://www.ifc.org/

progress towards achieving the United Nations (UN) 2030 Agenda, and the UN's own 17 Advocacy Targets[29]. The UN 2030 agenda concerns seventeen (17) Sustainable Development Goals (SDGs)[30] to be realised by 2030 including lofty goals for all like 'No Poverty (SDG 1)', 'Zero Hunger' (SDG 2), 'Good Health and Well-Being' (SDG 3), 'Industry, Innovation and Infrastructure' (SDG 9), 'Partnerships to achieve the Goals' (SDG 17), etc.[31]. Digital Inclusion and Meaningful Universal Connectivity (ensuring no one will be left behind in the digital era) are frequently considered to be related to the UN's SDGs 1, 3, 4, 6, 8, 9, 10, 11, 16 and 17.

The indefatigable advocate known as the Internet Society with its goal to 'build, promote and defend the Internet' has been at it since its founding in 1992. The World Bank's (WB) vision is 'to create a world free of poverty on a liveable planet' whilst the International Finance Corporation[32] (IFC) 'advances economic development and improves the lives of people by encouraging the growth of the private sector in developing countries'.

In one way or another, all these august organisations would like to see the *universal digital inclusion challenge* realised. The unequivocal top-level vision

---

[29] https://www.sightsavers.org/policy-and-advocacy/global-goals
[30] https://sdgs.un.org/goals
[31] The 17 Sustainable Goals are as follows:
- Goal 1: No Poverty
- Goal 2: Zero Hunger
- Goal 3: Good Health and Well-being
- Goal 4: Quality Education
- Goal 5: Gender Equality
- Goal 6: Clean Water and Sanitation
- Goal 7: Affordable and Clean Energy
- Goal 8: Decent Work and Economic Growth
- Goal 9: Industry, Innovation and Infrastructure
- Goal 10: Reduced Inequality
- Goal 11: Sustainable Cities and Communities
- Goal 12: Responsible Consumption and Production
- Goal 13: Climate Action
- Goal 14: Life Below Water
- Goal 15: Life on Land
- Goal 16: Peace and Justice Strong Institutions
- Goal 17: Partnerships to achieve the Goals

[32] https://www.ifc.org

and goal of *the digital inclusion challenge* is to digitally-include everybody in the world:

- by connecting them using telecommunications networks (the ITU), or ensuring universal broadband connectivity for all – the vision of the ITU's Broadband Commission;
- by 'supporting and promoting the development of the Internet as a global technical infrastructure, a resource to enrich people's lives, and a force for good in society' – the vision of the Internet Society;
- by bridging the digital divide in all of the UN's 194 member countries – the goal of the ITU's WTDC series of meetings;
- by ensuring the Internet is a basic human right for all 194 UN member countries – a key goal of the United Nations;
- by 'building resilient infrastructure, promote inclusive and sustainable industrialisation and foster innovation' – SDG Goal 9 of the UN;
- by building, promoting and defending the Internet for all – the mission of the Internet Society;
- by ensuring most of the world is connected for their altruistic financial interests – a key mission of the Big Tech behemoths like FANGAM (i.e., Facebook/Meta, Apple, Netflix, Google, Amazon, Microsoft), GAMMA (i.e., Google, Apple, Microsoft, Meta, and Amazon), etc.[33];
- indeed, as the ITU notes[34], Information and Communication Technologies (ICTs) can help accelerate progress towards every single one of the 17 United Nations Sustainable Development Goals (SDGs) – further emphasising the need digital [ICT] inclusion.

However, it is *unarguable* today in early 2025 that *none* of these above universal digital inclusion-related goals, missions or vision statements would materialise in most (if not all) of the OECD[35]-defined 'developing countries'[36] by the UN's chosen 2030 Agenda year, i.e., in 5 years' time. And there are 143 countries on

---

[33] Also check acronyms like FAAAM or FAANG
[34] https://www.itu.int/en/mediacentre/backgrounders/Pages/icts-to-achieve-the-united-nations-sustainable-development-goals.aspx
[35] Organisation for Economic Co-operation and Development (OECD) - https://www.oecd.org/
[36] https://www.gov.uk/government/publications/countries-defined-as-developing-by-the-oecd/countries-defined-as-developing-by-the-oecd

this 'developing countries' list, i.e., close to 74% of all 194 UN member countries. The reasons are many and varied.

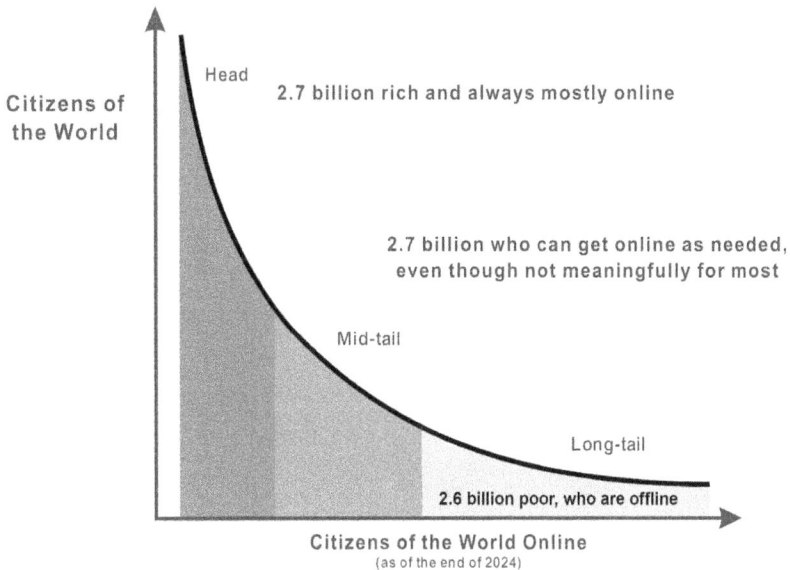

Figure 1.1 – The Long Tail Model[37] - the *head* "2.7 billion rich and always mostly online," the *mid-tail* " 2.7 billion who can get online as needed, even though not meaningfully for most" and *the tail* "2.6 billion poor[38] who are *offline* in 2025."

This book contends that all the esteemed aforementioned organizations, despite their universal digital inclusion visions and missions, exhibit limited *fit-for-purpose implementable solutions, commitment, sustained leadership, and management* necessary to actualise their ambitious goals. Furthermore, mere discussions at WTDC conferences will not resolve these challenges.

As the author of this book, I believe I have some [relatively deep] expertise on the universal digital inclusion challenge[39]. I hail from an OECD-defined

---

[37] Source: https://medium.com/@jasper_39476/how-will-long-tail-affect-the-leisure-branch-dec45338a5a
[38] 'Poor' as defined by World Bank and UN developmental metrics and nomenclatures, e.g. LDCs, LICs, LMICs, etc.
[39] I am the author of *Telecommunications, Media & Technology (TMT) for Developing Economies* (Nwana, 2014, 550 pages), *The Internet Value Chain & the Digital Economy:*

'developing' country (Cameroon) giving me first-hand grounding and 'feel' of the inclusion challenge, and I have been principally working with dozens of developing countries over the last decade and a half including in South East Asia, Caribbean, the Middle East and Sub-Saharan Africa.

In this book, *The Connectivity Crisis - Half the World Left Behind*, I propose some 'out of the box thinking', fit-for-purpose, implementable digital inclusion solutions that some UN member states could adopt. I outline the *commitment* necessary, and the form of *leadership* and *management* required to realise the lofty visions to truly strive for digital inclusion for all most – along with a monitorable *commitment* model to measure and evaluate them. These would help accelerate the realisation of the digital inclusion for all. My digital inclusion solution comes from the realities of developing countries, ***not*** developed nations' solutions imposed on developing economies.

Let me return to Figure 1.1 and briefly unpack some of its key messages.

- *2.6 Billion Long Tail*: as noted right at the start of this chapter, according to the separate ITU Broadband Commission[40] and the International Telecommunications Union (ITU)[41], there is a stubborn 2.6 billion people as of the end of both 2023 and 2024 [as shown in Figure 1.1], respectively, that are still offline, a reduction from the estimated 2.7 billion people offline in 2022. So, an 'easier' 100M were taken online in the 2022 to 2023 year – 'easier' because the next 100M would typically and <u>progressively</u> be harder.
- *2.7 billion Mid-Tail*: I have a healthy and well-founded scepticism how many of the mid-tail "2.7 billion who can get online as needed" of Figure 1.1 are truly meaningfully connected – as I cover in the next section.
- *2.7 billion Head*: these mostly-rich are also most meaningfully connected.

At a constant rate of 100 million connections per year, connecting the entire 2.6 billion Long Tail would take 26 years, extending past 2050. This is of course assuming a constant population across all continents during these next 26 years

*Insight and Guidance on Digital Economy Policy and Regulation* (Nwana, 2022, 400 pages), and his most recent book on *Demystifying Regulation - A Practitioner's Guide: Theory, Methods and Practice* (Nwana, 2024, 450 pages).
[40] https://www.broadbandcommission.org/publication/state-of-broadband-2023/
[41] https://www.itu.int/en/mediacentre/Pages/PR-2024-11-27-facts-and-figures.aspx

and ignoring population growth. Twenty-six years blows the UN 2030 agenda out of the water because realising all of the seventeen SDGs require [digital] ICTs.

Furthermore, the 'constant population' assumption is patently nonsense given that we already know that the most unconnected continent – Africa – has the fastest growing population. The World Economic Forum (WEF) has long projected that, by 2050, two in every five children will be born in Africa[42]. Therefore, it makes sense that to digitally-connect all by 2030 is already *not* realistic from today in 2025.

The other inescapable reality is that 2.6 billion offline are mostly *the poor*. The ITU figures[43] show that of the 5.4 billion people worldwide that are accessing the Internet in 2023/24, 93% of people in wealthy nations are connected compared to *only* 27% in Low-Income Countries (LICs). And most of low-income countries' connections are made on *usage-based mobile* connections with quite high data costs. This entrenches the digital divide as well as existing inequalities and hinders economic growth.

So, SMART[44] 'out of the box' solutions, thinking and enabling policy/regulatory environments are direly needed – *largely aimed at low-income countries*. Therefore, I hope that the august organisations, mentioned in this section, do not *only* take such proposed thinkings and insights in this book seriously, *but positively 'adopt' and own them*. Otherwise, their well-meaning intentions and visions will not even happen by 2050, yet alone 2030.

## 1.2 The 'Mid-tail' 2.7 Billion and 'Meaningful' Connectivity

I ask the following question semi-rhetorically: does the ITU *genuinely* believe that the bulk of the 'Mid-tail' 2.7 billion (see Figure 1.1) are 'meaningfully' connected? My clear answer is No!

---

[42] https://www.weforum.org/agenda/2020/01/the-children-s-continent/
[43] https://www.itu.int/en/mediacentre/Pages/PR-2024-11-27-facts-and-figures.aspx
[44] Specific, Measurable, Actionable, Realistic & Timely

I am concerned that the ITU [and the Broadband Commission] are deliberately using a definition of being connected to the Internet [online] which significantly 'increases' the mid-tail 2.7 billion number in Figure 1.1 – even though many of these 2.7 billion are not 'meaningfully connected'. This is because the ITU's approach and manual for measuring ICT and Internet access use[45] by households and individuals access asks [*verbatim*] questions like the following (the ITU collects annual telecommunication/ICT infrastructure and access data using both short and long questionnaires)[46]:

- "What types of goods or services did you buy or order over the Internet for private use *in the last 3 months*"?
- "How did you pay for the goods or services you bought over the Internet for private use *in the last 3 months*"?
- "How did you receive the goods or services you bought over the Internet for private use *in the last 3 months*"?
- What are the reasons why you did not purchase goods or services over the Internet for private use *in the last 3 months*"?

My point is that accessing the Internet *once* in 90 days ["in the last 3 months"] is an incredibly <u>minimal</u> measure that does not necessarily, in my opinion, reflect *meaningful connectivity*. It certainly is not as meaningful as that of the average UK adult having spent 56 days online in 2023 – which I note right in my Preface at the beginning of this book! I note that young UK adults spend – by my math – almost 70 days a year online in the UK. By all definitions, this is real meaningful connectivity.

Meaningful connectivity – by my definition - implies <u>regular, dependable, and effective</u> access to the Internet - one that enables the users to benefit from online resources and services <u>anytime, anyplace and anywhere</u>. This hardly translates to once in three months.

So, the ITU survey's questions about accessing the Internet at least once in the last 90 days only indicates the <u>most basic of basic access</u> – if indeed this even

---

[45] *Manual for measuring ICT access and use by households and individuals – 2020 Edition* - https://www.itu.int/dms_pub/itu-d/opb/ind/D-IND-ITCMEAS-2020-PDF-E.pdf
[46] https://www.itu.int/dms_pub/itu-d/opb/ind/D-IND-ITCMEAS-2020-PDF-E.pdf

suffices as 'basic access'. It does not capture the realities and nuances of meaningful connectivity which depends on factors such as frequency of use, quality of connection, purpose of use, digital skills, and affordability. These are all essential to understand the true extent of Internet engagement.

Therefore, I recommend the ITU constructs a more appropriate 'meaningful connectivity' question that provides a more nuanced and comprehensive picture by incorporating questions in the survey and asks about:

- Frequency of Internet use (e.g., daily, weekly, monthly, 90 days, etc.)
- Quality and speed of the Internet connection
- Activity types performed online (e.g., e-Commerce, social media and socialising, education, work, health, agriculture, etc.)
- Affordability of accessing the Internet
- Subjective Confidence in their digital skills (e.g., Basic, Low-Intermediate, Intermediate, Good, and Very Good)

It is only through considering such factors as the above that the ITU (and the rest of us) can better understand how users are using the Internet and whether they have *meaningful connectivity* that enhances their lives (living, working and playing).

I am confident that the meaningfully-connected 'mid tail' of users connected to the Internet in 2025 (see Figure 1.1) – whatever the ITU settles on as their definition for meaningful connectivity – is likely to be *only a fraction* of the 2.7 billion number, more likely even less than 40%. The ITU should prove me wrong. I return to this issue in Chapter 6 and later in Part III of this book.

## *1.3   The Pareto Principle and Universal Digital Inclusion*

However, notwithstanding the meaningful connectivity narrative introduction of the previous section, most of the real efforts to digitally-include the ITU's long tail 2.6 billion unconnected is still yet to come, i.e. *still ahead of us*.

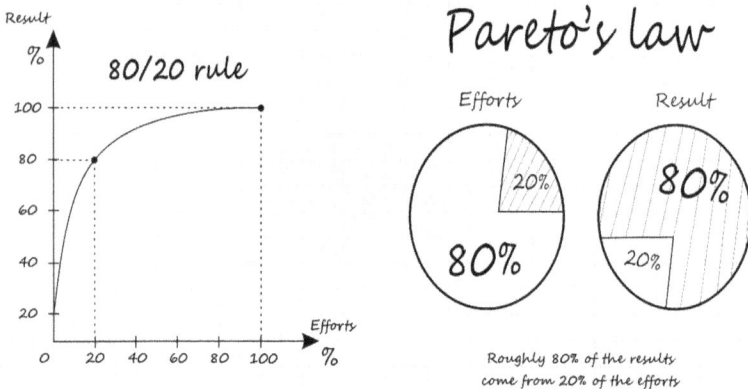

Figure 1.2 – The Pareto Principle (Adapted from Thomas Vato[47])

I already note earlier that this prediction is clear from the Pareto principle. The original Pareto principle as illustrated in Figure 1.2 is often referred to as the 80/20 rule. This *law of imbalance* was first realised and made famous in 1896 by Italian economist Vilfredo Pareto[48] when he realised the 'imbalance' that approximately "80% of the land in Italy was owned by only 20% of the population". He quickly realised that this 80/20 rule seemed to be a *universal heuristic* that could be applied to practically all aspects of life.

I have already acknowledged earlier that the Pareto principle is *not* scientific QED[49], but a rule of thumb that seems to always hold. Borrowing from Jānis Gulbis [50] with some additions from me, consider the following:

- 80% of results come from 20% of the effort (see Figure 1.2).
- 80% of outputs are produced by 20% of inputs.

[47] https://medium.datadriveninvestor.com/pareto-principle-the-framework-of-efficiency-e84f25a0dccc
[48] https://www.britannica.com/biography/Vilfredo-Pareto
[49] Q.E.D. stands for the Latin phrase *quod erat demonstrandum,* which translates to "that which was to be demonstrated."
[50] https://eazybi.com/blog/the-80-20-rule

- 80% of the world's riches are concentrated in 20% of developed countries.
- 80% of warehouse stock comes from 20% of suppliers.
- 80% of the world's food is grown in 20% of the world's countries.
- 20% of your friends will consume 80% of the beer at a party.
- 80% of the time, you visit only 20% of your friends or relatives.
- 80% of the taxes collected by Government comes from 20% of the tax paying base.
- 80% of the use of your high-end smartphone come from 20% of its functions.
- 20% of the features of your expensive high-end car is used 80% of the time.
- 20% of the clothes you own are worn 80% of the time.
- *80% of readers will read only 20% of this book.*
- Etc.

### 1.3.1 The Pareto Principle and the Effort to Lift the Unconnected 2.6 Billion Online

Permit me to address a question you may be wondering about so far: *what has the Pareto principle really got to do with the universal digital inclusion challenge?* I also want to answer the question in this section: *exactly how much effort does the Pareto principle predict still needs to be expended to lift the 2.6 billion offline into connectivity?* Particularly given that the digital inclusion challenge is one that largely afflicts low-income countries.

Well, I surmise and posit there is a most important insight (or prediction) to be gleaned from it (that I have noted earlier already) – but one I want to elaborate upon some more *to quantify the true measure of the effort required.*

Given that so far in 2024/25 there are still 32.5% of the world (or 2.6 billion people) offline, we have not even arrived at the 20% digitally excluded mark yet, since only 67.5% are connected. As I note in the previous section, this already tells us that more than 80% of the effort left is still ahead of us today in 2025. Figure 1.2 further illustrates this.

However, we can be even more specific, i.e., we can even do a bit of Pareto math to tell us how much 'effort' has gone into connecting the 67.5% digitally included to date, starting with the classic Pareto 80/20 rule. How do we do this? Therefore, permit me to divert into a bit of Pareto math because we will come back to this in a future chapter of this book. Do not be scared about the mathematics here – you need *not* be math-phobic.

| Power | 80/20 Rule applied to taxpayers | How the Math works (where the 'effort' in Figure 1.2 derives from) | Number of Taxpayers |
|-------|-------|-------|-------|
| 1 | 10000 | 0.2 x 10000 - paying 80% (0.8) of all taxes | 2000 |
| 2 | 2000 | $(0.2 \times 0.2)$ x 10000, i.e. $(0.2)^2$ x 1000 - paying $(0.8)^2$ of all taxes, 64%. | 400 |
| 3 | 400 | $(0.2)^3$ x 10000 - paying $(0.8)^3$ of all taxes, 51.2%. | 80 |
| 4 | 80 | $(0.2)^4$ x 10000 - paying $(0.8)^4$ of all taxes, 41%. | 16 |
| 5 | 16 | $(0.2)^5$ x 10000 - paying $(0.8)^5$ of all taxes, 32.7%. | 3.2 |

Table 1.1 – Illustrating the use of the 80/20 Rule with Taxes (Source: Author)

We can start with the scenario rule which states that 80% of the taxes collected by Government comes from 20% of the tax paying base. Let us start by assuming a tax base of 10,000 taxpayers. 80% of the taxes collected would be raised from 20% of the tax base, i.e., from 2000 taxpayers – as shown in Table 1.1.

However, this Pareto 80/20 rule also works *recursively or infinitely*. This is because even amongst the 2000 taxpayers, the 80/20 rule applies as well, and similarly amongst the 400 taxpayers as illustrated in Table 1.1. The 80/20 Pareto rule predicts those that are **truly** contributing to the 'effort' (see Figure 1.2):

- that just 80 taxpayers pay 51.2% of all taxes (see Row 3 of Table 1.1);
- just 16 taxpayers pay 41% (Row 4);
- and just above 3 taxpayers pay 32.7% (Row 5).

I hope Table 1.1 demystifies the 80/20 rule math.

So, to work out how much effort has been expended to deliver 67.5% of the world's population being digitally included as of 2025, this implies 32.75% are excluded, i.e., 0.3275.

Table 1.1 shows us the 'effort' level numbers – that 32.7% taxes are collected from just 3.2 taxpayers of the ten thousand. We have 32.75% digitally excluded today in 2025, which is remarkably close to the 32.7% *last-row* effort level in Table 1.1. So, we can use similar math in the last Row of Table 1.1 to work out the *effort level* to achieve or realise the inclusion of the remaining digitally excluded, i.e. $(1-0.3275)^5$, or $(0.675)^5$. If you use your calculator to do this, it will give you the result of 0.14.

So, in percentage terms, Pareto 80/20 predicts that <u>only</u> **14%** of the required effort has been expended to date to get to the 67.5% digitally connected to date in 2024/25. This means the bulk of the effort (86%) to connect the 2.6 billion digitally unconnected is still ahead of us!

The reader can verify my math too by just reading the effort level to date of the 80/20 curve of Figure 1.3. If you draw a line at the 67.5% (Y-axis) point to the 80/20 curve as shown on Figure 1.3, it depicts the 14% effort level (X-axis) to date from my math above.

The reason I include these other curves too is because, the reader may quibble and say 70/30 rule is more appropriate for my country, i.e., 70% of results come from 30% of the effort (adapting from Figure 1.2). If the reader chooses the 70/30 curve, Figure 1.3 still predicts that only circa 25% of the effort has been expended to connect the 67.5% digitally included today, leaving 75% much effort ahead of us. If you choose the 90/10 rule – then we are in big trouble because we have 95% plus effort to go.

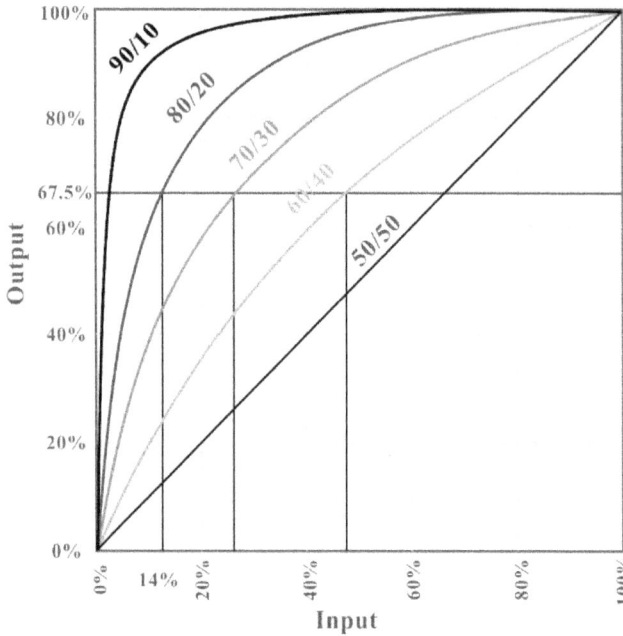

Figure 1.3 – The Pareto Distribution Curves (Source: Author, adapted from Jānis Gulbis [51])
– This figure also shows the curves for other rules: 90/10, 70/30, 60/40 and 50/50.

### 1.3.2  Key Learnings from the Pareto Principle

This is the bottom-line insight and learning from the Pareto principle.

Bottom line: the Pareto principle predicts or informs us that most of the *real digital inclusion 'hard graft' efforts are still ahead of us*. 86% more hard graft as shown in Figure 3.1 (Chapter 3). I wonder whether the UN with its UN agenda 2030 ever considered this reality. Even the GSMA projects a 73% global unique mobile subscriber number by 2030, up from a 68% penetration in 2022[52] – not 100%. The reader would note or learn further on in this book that Pareto applies to the connectivity crisis largely due to usage gap, demand-side challenges (starting from Section 2.5).

---

[51] https://eazybi.com/blog/the-80-20-rule
[52]    https://www.gsma.com/mobileeconomy/wp-content/uploads/2023/03/270223-The-Mobile-Economy-2023.pdf

## 1.4 The Need for 'Out-of-the-Box' Thinking and Solutions

The 86% load effort ahead of us needs 'out-of-the-box' thinking and solutions. Using the 80/20 curve from Figure 1.3, we now know from the previous section that 86% of the effort to connect the rest of the 2.6 billion digitally-excluded is *still to be expended* – as depicted in Figure 3.1.

This dawning that 86% effort levels to digitally-connect the 2.6 billion tail are still ahead of us may lead to some of these august organisations to 'throw in the towel'. I hope they do not. I argue in this book that we need to rethink our approaches to connect this 2.6 billion people because the approach or approaches that connected the 5.4 billion people – albeit a good percentage of them not meaningfully – *will not work* to connect these unconnected billions. So, I cover some of the characteristics of the sort of approaches needed in Chapter 3.

## 1.5 The Organisations and Audience of this Book

This book is *unashamedly and principally* targeted at developing nations, specifically the 143 OECD-defined 'developing countries'[53] – and of course their ICT leaders. Alternatively, the 137 nations defined by the World Bank as 'developing economies' are the target of the interventions of this book. The 137 countries are all included in the OECD's list of 143 countries.

For 2024, the World Bank classified countries and territories whose GNI[54] was $13,205 or higher as High Income Countries (HICs) economies. Anything below this number is considered a *developing country*, though World Bank prefers the terms Upper-Middle Income Countries (UMICs), Lower-Middle Income Countries (LMICs), and Low Income Countries (LICs). This classification helps provide a framework and guidance for understanding the economic development levels/needs of different countries, guiding necessary policies, investments, and international aid.

---

[53] Countries defined as developing by the OECD - GOV.UK (www.gov.uk)
[54] Investopedia defines Gross National Income (GNI) as the total amount of money earned by a nation's people and businesses. It is used to measure and track a nation's wealth from year to year.

In 2024, the World Bank-stipulated the following GNI per capita thresholds[55] (these thresholds are adjusted annually for inflation using the Special Drawing Rights (SDR) deflator to keep them fixed in real terms):

- HICs: GNI per capita of $13,205 or higher
- UMICs: GNI per capita between $4,046 and $12,535
- LMICs: GNI per capita between $1,046 and $4,045.
- LICs: GNI per capita below $1,046.

In 2022, only 85 of the 217 countries and territories were assessed by World Bank qualified as HICs (i.e. 39%), whilst 132 qualified as developing economies/countries[56]. In summary, using World Bank-speak, the LMICs and LICs are the target countries for this book. It is certainly *not* aimed at developed countries or HICs – whose citizens are almost universally digitally-included already – and mostly meaningfully connected too.

More elaborately, I have five audience categories in mind for this book:

1. *Leaders of Developing Economies – Presidents, Prime Ministers, Senior ICT Ministers, ICT Policy Makers and Regulators in Developing Countries [OECD]*: the august organisations mentioned above largely and deeply influence these groups of Presidents, Prime Ministers, senior ministers, policy makers, regulators and others in developing economies. I urge and plead with these groups of stakeholders that the developed nations-proven solutions do **not** typically translate to developing economies. Therefore, they truly should be driven by their local evidence on the ground, rather than proposals from London, Paris, Washington, Silicon Valley – San Francisco, Shanghai, Geneva, Canberra or wherever in the developed world universe. These stakeholders [presidents, prime ministers, senior ICT policy makers and regulators] should have open minds on 'lever' and 'pulley'-type solutions that would deliver maximum outcomes to developing economies/countries – and indeed, seek to develop such locally-inspired solutions.

---

[55] https://blogs.worldbank.org/en/opendata/world-bank-country-classifications-by-income-level-for-2024-2025
[56] https://blogs.worldbank.org/en/opendata/world-bank-country-classifications-by-income-level-for-2024-2025

For example, M-Pesa (cf. Section 5.1) is a uniquely Kenyan/Sub-Saharan Africa solution to a problem which is uniquely Kenyan and sub-Sahara African in nature. These stakeholders must ask more searching questions on universal digital inclusion and build these into out-of-the-box policies and regulations.

So, I believe there is no substitute to *inspired and quality leadership* from the top-most leaders in every developing economy to address the digital inclusion challenge. Unquestionably, there is no substitute for leadership which provides the Vision, Mission and Management to realise priorities. Rwanda's President Paul Kagame's leadership in his country since he started leading the country in 2000 after the Rwanda genocide of 1994[57] speaks for itself. Like him or loathe him, he has led Rwanda from the dark ages of 1993/94 into building a sort of 'miracle' in his country, relying initially on mostly uneducated guerrilla fighters and a handful of ill-trained cadres. This is a leader who regularly reviews the ICT projects in his country. I appeal to leaders of developing countries to lead – personally – on digital inclusion as President Kagame has been doing – including being a co-Founder of the Broadband Commission. However, as I point out in Table 6.3 (Chapter 6) later in this book, Rwanda still has a long way to go in addressing its universal and meaningful connectivity challenges.

2. *The august organisations mentioned in this chapter (Broadband Commission, the myriad Aid and Developmental Agencies (USAID[58]), UK Aid, Australia Aid, etc.[59]), the ITU and the UN in particular)*: they include

---

[57] https://www.britannica.com/event/Rwanda-genocide-of-1994

[58] U.S. Agency for International Development (USAID) – currently and sadly as of Q1/Q2 2025 being 'dismantled' by the Trump 2.0 Administration.

[59] A longer more comprehensive (non-exhaustive) list includes (in no particular order):
1. African Development Bank (AfDB)
2. World Bank – International Development Agency (IDA)
3. Inter-American Development Bank (IaDB)
4. Asian Development Bank (ADB)
5. US Millennium Challenge Corporation (MCC)
6. U.S. Agency for International Development (USAID)
7. United Nations Children's Fund (UNICEF)
8. United Nations Development Programme (UNDP)
9. Bundesministerium für wirtschaftliche Zusammenarbeit (BMZ-GIZ). Deutsche Gesellschaft für Internationale Zusammenarbeit (GIZ) is the German development agency that implements projects on behalf of the German government, including BMZ.

the ITU in general, the ITU's Telecoms Development Sector (ITU-D) in particular[60], the Broadband Commission, the Internet Society, the WTDC Conference theme organisers, the UN 2030 Agenda-setting team, the United Nations in general with its seventeen SDGs, the GSMA[61], the World Bank, the IFC and Big Tech (FANGAM/GAMMA) companies like Google, Meta/Facebook, Amazon, etc. These organisations – whether they believe it or not – individually and collectively have an outsized and clearly visible influence on developing economies' Presidents, Prime Ministers, Senior ICT Ministers, policy makers and regulators.

---

10. China - Ministry of Commerce of the People's Republic of China (MOFCOM)
11. Korea International Cooperation Agency (KOICA)
12. European Commission (EC) Directorate-General for International Partnerships (INTPA) - a department within the EC responsible for formulating the EU's international partnership and development policy. Its mission is to reduce poverty, ensure sustainable development.
13. United Kingdom - Foreign, Commonwealth & Development Office (FCDO)
14. France - Agence Française de Développement (AFD)
15. Canada - Global Affairs Canada (GAC)
16. Sweden - Swedish International Development Cooperation Agency
17. Japan - Japan International Cooperation Agency (JICA)
18. New Zealand - Ministry of Foreign Affairs and Trade (MFAT)
19. Gates Foundation
20. European Bank for Reconstruction and Development (EBRD)
21. European Investment Bank (EIB)
22. Italy - Italian Agency for Development Cooperation (AICS)
23. World Bank/International Finance Corporation (IFC)
24. Ireland – Irish Aid
25. Switzerland - Swiss Agency for Development and Cooperation (SDC)
26. Australia - Department of Foreign Affairs and Trade (DFAT)
27. Spain - Spanish Agency for International Development Cooperation (AECID), or Agencia Española de Cooperación Internacional para el Desarrollo
28. Saudi Arabia – Kingdom of Saudi Arabia Relief (KSRelief)
29. IDB Invest - IDB Invest is the private sector arm of the Inter-American Development Bank (IDB)
30. Turkey - Turkish Cooperation and Coordination Agency (TIKA)
31. UAE - Ministry of Foreign Affairs and International Cooperation (MOFAIC)

[60] https://www.itu.int/en/ITU-D/Pages/About.aspx
[61] https://www.gsma.com/

I know from personal professional experience that *developed countries'* policy makers, regulators and operators *can* adopt their own local-market, evidenced-driven policies and regulations – irrespective of what these august organisations advocate for. I know this personally because I was a senior regulator in the UK at Ofcom[62] leading on such regulatory decisions. However, I see regularly how swayed *developing economies'* policy makers and regulators are by these organisations. Therefore, I see it as paramount that they [these august organisations like the ITU, UN, Broadband Commission, Aid and Developmental Agencies, GAMMA, etc.] buy in into such insights and 'out-of-the-box' solutions and thinkings as I advocate in this book. Otherwise, I would be metaphorically pushing water uphill inside developing countries with such ideas.

3. *Telecommunications and broadcast network organisations*: these include satellite, Wi-Fi Players, ISPs, FWA providers[63] and more. Let me note the following.

   **Satellite Operators**: they really need to hear and learn this. From a developing economies standpoint (sub-Saharan Africa, South East Asia, some Latin America, Caribbean islands, SIDs, etc.), they [Satellite Operators] have promised so much over the last 30 to 50 years to universally digitally-connect all, but – sadly – largely failed to deliver. Had developing countries been waiting for satellite as the universal connectivity solution, we would still be in the dark ages of the 1970s.

   As Figure 1.4 hopefully clarifies, Universal Service is any service that the Government expects to be a*vailable, affordable* and *accessible* throughout the population. Therefore, universal service has two key aspects to it. First, it means the service must be made <u>available</u> to all <u>X</u>[64], perhaps within a given

---

[62] https://www.ofcom.org.uk/

[63] Fixed Wireless Access (FWA) prioritises connectivity for fixed locations in contrast to mobile networks designed for on-the-go connectivity. So, FWA provides internet to static location likes homes, offices, or businesses through a wireless link. Customers install an antenna or receiver at their premises to connect to a nearby mobile tower which could be 4G- or 5G-based, or even based on other wireless technologies like satellite, TV Whitespace or even Wi-Fi.

[64] Governments would typically define what "X"s are: e.g. districts, homes, schools, health centres, etc.

area or of the entire population. Second it should be made <u>available</u> at a uniform and <u>affordable</u> price.

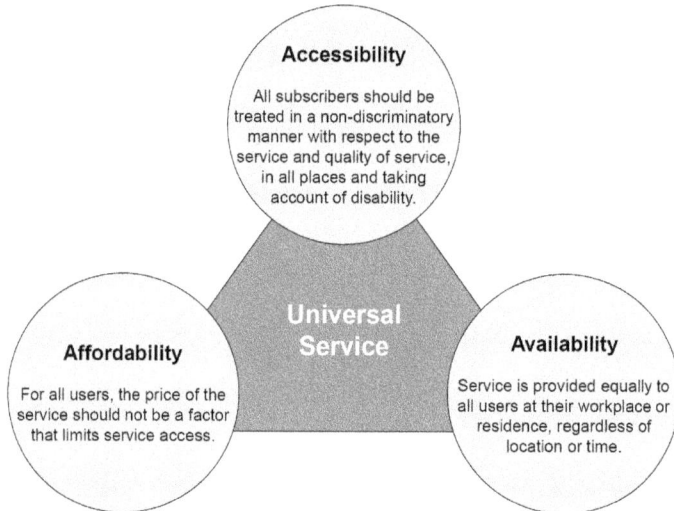

**Accessibility**

All subscribers should be treated in a non-discriminatory manner with respect to the service and quality of service, in all places and taking account of disability.

**Universal Service**

**Affordability**

For all users, the price of the service should not be a factor that limits service access.

**Availability**

Service is provided equally to all users at their workplace or residence, regardless of location or time.

Figure 1.4 – What is Universal Service? (Source: Dr Charles Jenne[65])

Whilst connectivity satellite solutions are evidently [theoretically] available, they have failed to be accessible and affordable, as per Figure 1.4. The satellite industry is going through a true 'mini-revolution' with all the LEO-constellation satellites being launched by the likes of Starlink, Amazon's Project Kuiper, Telesat's Lightspeed[66] and others. However, and sadly, developing economies are already beginning to observe the age-old challenges with satellite and universal access – they are not available (everywhere), not accessible and evidently not affordable, e.g. see Forbes Home analysis of Starlink's Plans, Prices and Speeds[67]. Satellite providers must think out of the box on accessibility and affordability in particular.

***MNOs***: mobile network operators have changed the lives of billions in developing economies – this is unequivocal. As I mention earlier, according to the respected GSMA's 2023 Mobile Economy Report[68], there were 5.4

---

[65] Dr Charles Jenne is a close collaborator and colleague of the author at Cenerva Ltd.
[66] https://www.satelliteinternet.com/resources/what-is-low-earth-orbit-satellite-internet/
[67] https://www.forbes.com/home-improvement/internet/starlink-internet-review/
[68] https://www.gsma.com/mobileeconomy/wp-content/uploads/2023/03/270223-The-Mobile-Economy-2023.pdf

billion *unique* mobile subscribers globally in 2022 (i.e., 67.5% world penetration), with the GSMA projecting a 73% penetration by 2030. According to the ITU, an estimated 5.5 billion unique people were online in 2024, an increase of 227 million individuals based on revised estimates for 2023[69].

Clearly, even the mobile industry is stalling with the connectivity challenge. There has been extraordinarily little out-of-the-box 'lever' and 'pulley' type mobile solutions of late to accelerate this 73% GSMA 2030 goal. It is worrying that MNOs are now the new "fixed incumbents" of the pre-1990s, concentrating on "sweating and milking" their network assets in many developing countries. With such complacency setting in, policy makers and regulators must enable other newer players to enter the digital inclusion space – as discussed next in the case of Brazil.

***ISPs, Wi-Fi and FWA Players***: in many (if not most) developing countries, these players are *de minimis* to providing universal connectivity solutions, i.e., so minor as to be disregarded compared to the substantial players above. This has mostly been driven by Government policies and regulations. It should not necessarily be this way. Some developing economies are beginning to do things differently. In 2021, Brazilian telecoms regulator ANATEL enabled Brazil to become one of the first countries to open up unlicensed access in the 5925 – 7125 MHz band using new standards like Wi-Fi 6E in this 6GHz band. This has greatly advanced the country's connectivity goals and helped support over 20,000 ISPs to provision more reliable W-Fi using the 6 GHz band.

Today in Brazil, ISPs lead in growth and market share, providing Internet connectivity for more than 50% of nearly 45 million fixed access points, which are mostly enabled via fibre to the home (FTTH)[70]. How many developing economies have similarly promoted such different ecosystem players to address the universal digital inclusion challenge?

---

[69] https://www.itu.int/en/mediacentre/Pages/PR-2024-11-27-facts-and-figures.aspx
[70] https://www.dynamicspectrumalliance.org/wp-content/uploads/2019/06/AFCPressAnnouncementBrazil-1.pdf

***Digital Terrestrial TV (DTT) Broadcast Networks Players***: DTT networks are dying on their feet in most developing economies. Erstwhile only TV-providing cable companies in the USA - post the USA's 1996 Telecoms law – transformed themselves with the introduction of the DOCSIS standard (in the USA) into fierce competitors to the then-traditional telecommunications businesses, and this is still the case today in 2024. I believe that even traditional DTT networks – with some out-of-the-box imagination – could be key networks to enable digital inclusion for all through the digital TVs in homes.

4. *Suppliers Integrating Standards and other Solutions* (e.g., solar suppliers, telco, broadcast and other suppliers): as I cover in Part II of this book, I think that suppliers can evolve out-of-the-box solutions to the digital inclusion challenge through integration of their solar, broadcast, telecoms and other standards and solutions, including even finance solutions. However, the developing countries *demand* for such new solutions *need to be aggregated to make it worth the efforts of these suppliers to engage in such integrations.*

5. *Students and Entrepreneurs*: there are other important audiences of this book like university professors and lecturers, relevant digital inclusion NGOs, policy think tanks, development agencies, etc. I do not want to minimise this miscellaneous group as others who could benefit from my viewpoints in this book – at least, that is my hope. However, I particularly want to highlight students and entrepreneurs, particularly in developing countries. I believe they are the key to addressing the digital inclusion challenge. I encourage them to look at the problem with fresh eyes, and from different perspectives as I poorly start to define in this book. They should not be shackled with current solutions. Entrepreneur Steve Jobs changed the world in 1997 with his iPhone Vision. He envisioned three things in one: *a widescreen iPod with touch controls; a revolutionary mobile phone; and a breakthrough Internet communications device.* An iPod, a phone, and an Internet communicator – not as three separate devices, but as one device which he called the iPhone. Steve Jobs reinvented the phone! I just feel – some smart students and entrepreneurs will reinvent the connectivity to the Internet too.

## 1.6 Book Structure and Reading Guide

### 1.6.1 Breakdown of the Book

Following is a breakdown of the rest of the book. It consists of **three core Parts** with ten chapters.

I.  *Introduction, Realities of the Universal Digital Inclusion Challenges and Characteristics required of Out-of-the-Box approaches to its resolution*

**Part I** includes *this* introductory **Chapter 1** on the introduction to the universal digital inclusion and meaningful connectivity challenges. It also draws from Pareto to explain why we are still at the foothills to realising these challenges. It also includes Chapter 2 and Chapter 3.

**Chapter 2** describes the realities of the digital inclusion challenge as faced on the ground using the experience of some real countries as 'case examples' to illustrate and 'ground' the universal digital inclusion challenge. Many a time, the challenge of universal digital inclusion comes across *as much simpler* when sitting in Board rooms Geneva, New York, Washington, London or Paris.

**Chapter 3** builds on Chapter 1 and 2 to propose six key sets of characteristics required of approaches and [network] solutions to lessen the load effort of 'lifting' the long tail 2.6 billion into online connectivity.

II.  *An Innovative Solution and Some Cautionary Warnings*

**Part II** consists of three chapters: Chapters 4, 5 and 6.

**Chapter 4** – Civilisation in a Box – describes one proposed solution that encompasses many of the characteristics proposed in Chapter 3, and that implements key aspects drawn from the summarised challenges to connecting the Tail 2.6 billion from Chapters 1, 2 and 3.

**Chapter 5** makes the case for democratising and localising connectivity responsibility. It argues for Big Governments 'getting out of the way' in LDC,

LIC and LMIC economies and empowering locals to take (co)responsibility of perennial issues in their communities, including lack of digital connectivity.

**Chapter 6** may be considered controversial by some. It warns that the telecoms industry is increasingly giving up connecting most of the tail 2.6 billion not connected to the Internet. As evidence, it points out that it has already failed some circa 1.3 billion globally who could not *regularly* make a basic 2G voice call in 2024. It concludes that strong regulatory intervention is very necessary going forward.

III. *Strategy, Policy, Regulatory, Collaboration & Leadership Levers*

**Part III** consists of four chapters: Chapter 7, 8, 9 and 10

**Chapter 7** on Strategy and Collaboration addresses the broad issues of *individual country* strategies and the *inter-country* collaboration strategies - that would drive up the connectivity amongst the 2.6 billion of people *not* connected to the Internet as of early 2025 and help meaningfully connect 2 billion others.

**Chapter 8** makes Policy and Regulatory recommendations that would/may mitigate the connectivity digital divide amongst the unconnected 2.6 billion to the Internet.

**Chapter 9** on Leadership is the chapter that gives this book its sub-title: "Expert Insights for Global Institutions and Developing Nations Committed to Achieving Universal Digital Inclusion". It overviews many categories of leadership that are all involved in some way in bridging the connectivity digital divide across the globe, not least because it gets quite confusing.

**Chapter 10** summarises the book.

### 1.6.2 How to Read the Book

Regarding how you may choose to read the book, permit me to recommend you read my *Preface* first, then read this Chapter (Chapter 1) and then read Chapter 10 on *The Book Summarised*. These should give you the measure of the book. Thereafter, you can read the chapters in any order you like – though sequentially would be better.

Frankly what matters more is for you to be inspired enough to read most or all the chapters in the book in whatever order you choose. And even better, to *act* on what you read in these pages.

# 2 Digital Inclusion Realities in Low-Income Countries

At the start of Chapter 1, I note the ITU Secretary-General (SG), Doreen Bogdan-Martin commenting on the ITU's Facts and Figures 2024 report[71] as "a tale of two digital realities between high-income and low-income countries". The inescapable reality in 2025 is that ITU figures[72] show that 93% of people in wealthy nations are connecting compared to only 27% in Low-Income Countries. And connection in low-income countries is largely *usage-based* and quite expensive – hardly meeting any definition of 'meaningful connectivity'. For evidence, according to the ITU[73], in 2023, a 10 GB mobile data plan in low-income countries could consume up to 25% of a person's monthly income, compared to just 1–2% in high-income countries. The price gap for such plans is 36 times larger between low- and high-income economies. This is a staggering disparity, especially when you consider that 10 GB is a modest amount for modern Internet use. So, it is important that the digital inclusion challenges in low-income countries are better and truly appreciated.

## 2.1 Universal Digital Inclusion & UN's SDGs

Many a time I feel the well-meaning digital inclusion framers – based in New York, Geneva, London, Washington D. C., etc. – of goals like the seventeen UN SDGs listed in the last chapter had zero to *minimum idea* of the realities on the ground in developing countries, when they sat down and proposed *UN Agenda 2030* back in 2015. As I also argue in the previous chapter the Pareto principle would have predicted to them (and us all) that to realise these SDGs in fifteen years – for most developing economies/low-income countries – was completely implausible.

---

[71] *Ibid.*
[72] https://www.itu.int/en/mediacentre/Pages/PR-2024-11-27-facts-and-figures.aspx
[73] https://www.itu.int/en/ITU-D/Statistics/Documents/publications/prices2023/ICTPriceBrief2023.pdf

### 2.1.1 Stretch Digital Inclusion Targets vs. Patently Impossible Ones

I genuinely *do understand* the case for 'stretch targets', but *not patently impossible ones*. I do understand that such aspirational goals may focus minds in developing countries, but I do not see much evidence of these in practically all the ten to twenty developing economies I visit annually.

Conversely, I am always equally aghast and bemused when ministers of developing economies' member states go to New York and Geneva and sign up their countries to these goals – presumably, knowing fully well that they have no hope of realising them by 2030, or whatever the dates are. I ask myself 'what is the point'?

Such declarations do not only happen in New York or Geneva. Consider the following 2012 Indaba Declaration in Cape Town, South Africa:

> *"We, the Ministers responsible for Information Communication Technologies (ICTs) in our respective countries in Africa, assembled in Cape Town from the 4th–7th of June 2012 for the Inaugural ICT Indaba[74]...hereby declare our common desire and commitment to eradicate the barriers of poverty through the promotion and use of enabling ICTs to build and foster a people-centred knowledge-based economy in Africa....We declare access to broadband communication as a basic human right in Africa and commit to increasing broadband penetration to approximately **80 per cent of the population by 2020**. This common vision draws its basis from the positive impact exerted on economic growth through increasing Accessibility, Affordability, and Availability to broadband by all."[75]*

Impressive words, promises and targets like 80% broadband penetration "of the [African] population" from the Cape Town Indaba Declaration of 2012! Today

---

[74] "Indaba" is a Zulu word for a council or meeting of indigenous peoples of southern Africa who meet to discuss an important matter.

[75] Inaugural ICT Indaba: African ICT Ministerial Declaration, June 7, 2012. In June 2012, African ministers for ICTs attended the inaugural Indaba in South Africa event and set themselves the target of achieving broadband penetration of approximately 80 percent by 2020.
www.ictindaba.com/2012/images/ICT_Indaba_2012_African_ICT_Ministerial%20Declaration_07June2012.pdf

in early 2025, we are 5 years past 2020 already. It was at 57% globally in 2022 as shown in Figure 2.1 – yet alone ever realising 80% across Africa. As Figure 2.1 shows, Sub-Saharan Africa (SSA) was at 25% online-connected only, with the highest usage and coverage gaps of 59% and 15%, respectively. *Usage gaps* refer to populations that live within the footprint of a mobile broadband network but do not *use* mobile Internet. *Coverage gaps* refer to populations that do *not* live within the footprint of a mobile broadband network.

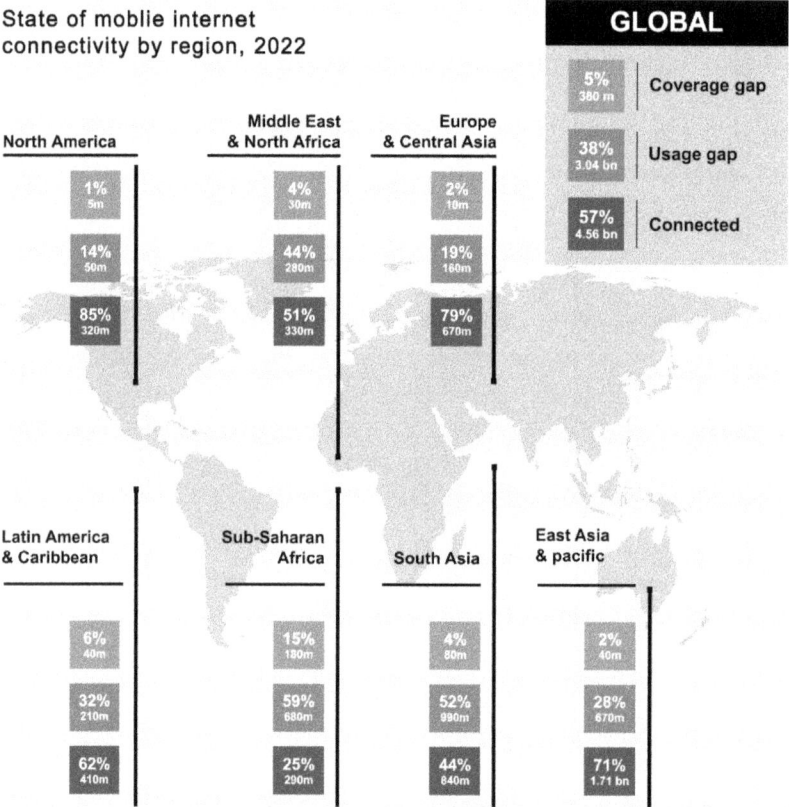

Figure 2.1 – State of Mobile Internet Connectivity by Region, 2022
Source: adapted from GSMA[76]

---

[76] https://www.gsma.com/r/wp-content/uploads/2023/10/The-State-of-Mobile-Internet-Connectivity-Report-2023.pdf

Hailing from Africa - and having worked (consulted) in many countries in SSA – I believe these GSMA figures are on the optimistic side, as they are self-reported by GSMA member MNO firms. 80% broadband connectivity in SSA by 2020 was never realistic as projected back in 2012 – so why did they bother?

What behaviours did this Indaba declaration of 2012 really change in SSA? I have seen almost zero evidence of true behaviour change, having worked closely with policy makers and regulators across more than 34 SSA countries over the past decade. Most *current* African ICT Ministers I meet have long forgotten this 2012 Indaba Declaration, and moved on to the UN 2030 Agenda SDGs which *they will also miss* by a *country mile in 2030.*

Other regions like South Asia, Latin America (LATAM) or the Caribbean have had their equivalent unique declarations too, like the 2012 Cape Town Indaba declaration over the last decade and a half. However, LATAM and the Caribbean are faring better than Sub-Saharan Africa and 'poor' Southeast Asia nations as regards online/broadband connectivity. As of January 2024, more than 80 percent of South Americans had access to the Internet, and some 70.2 percent of the Caribbean population. Mobile devices represented the largest shares of Internet access across Latin America throughout 2023[77].

### 2.1.2 The *Implausible* 2018 Broadband Commission Targets for 2025

Even the impressive ITU Broadband Commission is *not* immune from setting implausible broadband targets. In January 2018, at its Special Session during the Annual General Meeting of the World Economic Forum, the Broadband Commission for Sustainable Development set a total of seven broadband targets to support "Connecting the Other Half' of the world's population[78] – to be realised by 2025. They did this by updating and extending prior targets that were to be realised by 2020. The Commission set a goal of 75% of the world's population, including 65% of citizens of emerging markets and 35% of citizens in the least developed countries (LDCs), being connected to the Internet by 2025

---

[77] https://www.statista.com/topics/2432/internet-usage-in-latin-america/
[78] https://www.itu.int/en/mediacentre/Pages/2018-PR01.aspx

(albeit not necessary on a broadband connection and without a data consumption limit)[79]. To achieve its goals, the ITU Broadband Commission also set a target that entry-level broadband services should be made affordable in developing countries, at less than 2% of monthly gross national income per capita by 2025. One target also stipulated that, by 2025, all countries should have a *funded* National Broadband Plan or strategy or include broadband in their Universal Access and Service (UAS) definition. The ITU sensibly tracked data on broadband as defined as more than 256 kbps.

In its 2020 Report[80], the Broadband Commission rightly conceded that, at then-current rates of growth, it is/was unlikely that it will achieve its 2025 goals. For example, it acknowledged the 75%, 65% and 35% 2025 targets above were tracking well below where they should be to hit 2025. The Commission partly blamed Covid-19, but I argue that the targets were more than "ambitious and aspirational[81]" as the Commission termed them – they were simply implausible from the very beginning. They had already been simplified and the target dates extended from 2020 to 2025. Today, in 2025, some of the targets are not close to being met.

### 2.1.3 Key Learnings from Critiquing Digital Inclusion and Broadband Targets

My point of critiquing the UN 2030 SDGs, the ITU Broadband Commission's broadband targets, or the 2012 Cape Town Indaba Declaration, is twofold:

- first, to show their *implausibility* for many developing economies, hoping it will lead to better projections and target settings going forward.
- and secondly, to promote out-of-the-box Insights and Solutions to accelerate their realisations.

---

[79] *Ibid.*
[80] https://www.itu.int/itu-d/reports/broadbandcommission/2021/09/14/chapter-three/#advocacy_targets_block_pre
[81] *Ibid.*

For a start, it makes no logical sense to me that all the regions of the world shown in Figure 2.1 with their different mobile Internet connectivity data shown would be expected to reach 100% before or at the same time in 2030.

For both the UN SDG framers as well as the Ministers from developing countries who sign up to them, I believe it is important they better understand the on-the-ground realities of developing economies. The *specificities* of countries truly matter when setting digital inclusion goals. I think this is relegated many times to an afterthought when setting general digital inclusion goals thousands of miles away in Geneva, New York or London.

For this reason, I overview the challenging specificities of two developing countries next – one in Asia and one in Sub-Saharan Africa.

## 2.2 Universal Digital Inclusion Challenges for Papua New Guinea, South Asia

I am honoured to have worked with senior policy makers and regulators in Papua New Guinea (PNG) for some time and got to know and love the country. Figure 2.2 shows the map of PNG, a country to the Northeast of Australia.

It is not as apparent on this map, but PNG consists of circa 600 islands including the populated islands, making it an *archipelago*. The largest island is the Mainland on the Island of New Guinea (Figure 2.2) which is bordered by Indonesia. Other major islands are New Britain, New Ireland, Latangai, and Bougainville. PNG has a population of 10,443,815 people as of January 2024 according to Worldometers[82]. However, the PNG National Statistical Office (NSO)[83] estimates the 2021 population at 11,781,559.

Here are some further *current* (as in early 2025) specificities about PNG, because – as I note earlier – these *specificities* and *starting positions* matter to the universal digital inclusion challenge for PNG.

---

[82] Papua New Guinea Population (2024) - Worldometer (worldometers.info)
[83] https://www.nso.gov.pg/statistics/population/

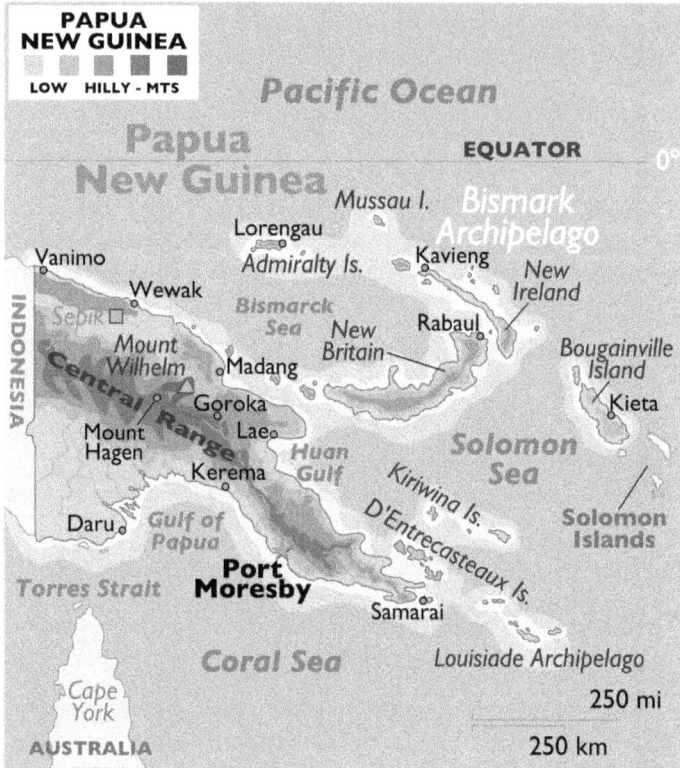

Figure 2.2 – Map of Papua New Guinea (Source: adapted from Shutterstock[84])

i. PNG is *an extremely rural and low population density country – a major challenge to digital inclusion*: only 12.3% of the population is urban (1,271,933 people in 2023). The total land area is 452,860 km$^2$ (174,850 sq. miles). Working from the population and the area gives a PNG population density of 23 per km$^2$ (59 people per mi$^2$). This is compared to a world population average density of 60.29 per km$^2$ according to Macrotrends[85]. For context, the population density of the United Kingdom is 275.27 per km$^2$, i.e. ten times that of PNG. In fact, the population density of England – one of the four countries that make up the UK – is circa 434 residents per km$^2$, some nineteen times that of PNG[86]. It is clearly much

[84] https://www.shutterstock.com/image-vector/papua-new-guinea-political-map-capital-741034618
[85] World Population Density 1950-2024 | MacroTrends
[86] https://www.gov.uk/government/publications/census-2021-first-results-england-and-wales/population-and-household-estimates-england-and-wales-census-2021

more efficient and economical to serve and digitally-include the citizens of UK and England than the citizens of PNG.

ii. PNG has *Connectivity levels amongst the lowest in the world with mobile penetration rates hovering around 40%*: this is according to PNG telecoms regulator itself NICTA in its 2021-2025 Corporate Plan[87]. According to the same Corporate Plan, household broadband penetration is/was less than 10% – no different today in early 2025: "despite increased network investment, most households are unable to receive broadband services to allow effective access to the internet due to limited access and the relatively high costs of internet. The challenge of connectivity still exists as household broadband penetration is less than 10%". According to datareportal.com[88], GSMA Intelligence suggests that there were a total of 3.74 million cellular mobile connections were active in Papua New Guinea in early 2023, with this figure equivalent to 36.5 percent of the total population. There were 3.29 million Internet users in Papua New Guinea at the start of 2023, when Internet penetration stood at 32.1 percent. The PNG Telecoms Regulator, the National Information and Communications Technology Authority (NICTA), reports that there were 3,538,595 mobile subscribers (i.e. SIM cards) by the close of 2023 in PNG, i.e. 30 percent of the PNG population[89].

The problem with these 36.5% and 32.1% or even the NICTA 30% mobile and Internet connectivity figures is that they significantly mask a big dual (or more) SIM problem, i.e., subscribers have two or more SIM cards each. The respected GSMA found in 2018 that PNG unique subscribers remain low compared to other countries in the region, with a mobile Internet penetration (unique subscribers) at *only* 11 per cent[90]. This was with a 2018 population at the time of 7.6M people compared to 11.7M people today

---

[87] https://www.nicta.gov.pg/about-us/corporate-plan/
[88] https://datareportal.com/reports/digital-2023-papua-new-guinea
[89] Source: PNG NICTA CEO Keynote Presentation, Digital Transformation Summit 2024, 2nd October 2024, Port Moresby, PNG -
https://www.ict.gov.pg/digitaltransformationsummit2024/
[90] https://www.gsma.com/solutions-and-impact/connectivity-for-good/mobile-for-development/wp-content/uploads/2019/03/Digital-Transformation-The-Role-of-Mobile-Technology-in-Papua-New-Guinea.pdf

according to the PNG National Statistical Office (NSO). I do not think this 11% unique subscriber level has increased that much compared to 2018 – it may have crept up a couple of percentage points or more – *but no more than 15-18% I estimate today in 2025*. The story is clear: PNG is one of the least digitally connected countries in the world. In 2023, the GSMA reported that 64% of PNG citizens did not have access to the Internet[91] at all (i.e., coverage gap in this case).

iii. PNG has *exceptionally Low ARPUs and Smartphone Connections in the Rural Areas where 87% of the Population Inhabit across 600 Islands*: this fact is extremely revealing of the universal connectivity challenge in PNG. According to the GSMA, Average Revenue Per User (ARPU[92]) in 2018 was circa USD $5.79 in urban areas. However, ARPU in rural and remote PNG was exceptionally low (USD ~$0.60–$0.90) with limited numbers of customers per site.

This obviously makes the business case for mobile tower deployments, operations and maintenance incredibly challenging without external subsidies. The percentage connection by smartphones 21.92%. Smartphone adoption largely mirrors the 30% *non-unique* Internet penetration across PNG, mostly in the urban areas and miniscule in the rural areas. So, for the non-urban areas, the combination of low population density, low incomes and inadequate access to electricity leads to very limited ability to access and use mobile phones.

iv. PNG has an extremely Limited *Formal Economy, Poverty, Low Literacy and Very low access to electricity*: the same GSMA report of 2018[93] reported that "PNG has a "two-tiered economy". One tier is occupied by about 70 per cent of the population who live at a subsistence level. Fewer than 250,000 people in this tier, or about one in 25, are formally employed. For this segment, the informal economy continues to be their dominant source of livelihood, with the bulk of economic activity taking place in

---

[91] https://www.gsma.com/solutions-and-impact/connectivity-for-good/mobile-for-development/blog/papua-new-guinea-challenges-and-opportunities-in-the-digital-worlds-of-displacement-affected-communities/
[92] Average Revenue Per User (ARPU) is a measure used primarily by mobile network operators, defined as the total revenue divided by the number of subscribers (calculated monthly)
[93] *Ibid.*

traditional subsistence farming. The other tier is the cash economy and includes agricultural products, mining, manufacturing and other industries and services. PNG has been described as having a "poverty of opportunity" or "lack of access to basic services, jobs, and education."

There was/is also a 37% illiteracy rate, meaning a *theoretical maximum* of 67% people would be able to use the Internet anyway, and less than 15% of the PNG population have reliable access to electricity. I strongly believe that these figures have not changed much today in 2025. Indeed, in 2022 the GSMA found that a camp in Western PNG was left without connectivity for six days due to difficulties delivering fuel to the mobile tower. The Western Province is one of the provinces most reliant on fuel delivery *by helicopter*[94].

*These [above] four specificities and starting positions of PNG are more than sufficient to paint the huge challenges to universal digital inclusion in Papua New Guinea.* It is true that mobile and Internet use in PNG is growing but is far from universal. In summary, GSMA Intelligence data indicates that only 35% of the population have a mobile phone and 21% have a mobile Internet connection across the country[95]. And that the country's digital ecosystem faces many and huge challenges, including physical barriers such as challenging terrains, limited intra-country road networks, natural hazards and low population density – making it prohibitively expensive for MNOs to build and maintain mobile infrastructure[96]. Furthermore, with 87% of the population living in rural areas and inhabiting 600 islands, it is even more cost prohibitive to connect them.

Additionally, GSMA notes that with more than 800 languages spoken in PNG and low levels of [digital] literacy, it is complicated and hard to create digital products and services and scale them across the country.

Thanks to the above four reasons, PNG's telecom coverage for the end of 2023 still leaves so much to be desired, as shown in Table 2.1.

---

[94] https://www.gsma.com/solutions-and-impact/connectivity-for-good/mobile-for-development/wp-content/uploads/2022/10/PAPUA-NEW-GUINEA_DW.pdf
[95] *Ibid.*
[96] *Ibid.*

| DEC 2023 | 2G | | 3G | | 4G | |
|---|---|---|---|---|---|---|
| | Population | % | Population | % | Population | % |
| Overall | 9,335,434 | 79% | 8,997,288 | 76% | 8,996,821 | 76% |

Table 2.1 - Mobile Coverage in PNG as of Dec 2023. Source: Adapted from [97].

Table 2.1 shows, that as of December 2023, a good 21% of PNG citizens are not even under any telecommunications and/or mobile coverage – even to make basic 2G voice calls to their fellow citizens.

Does the reader still believe now that PNG will attain UN Digital Agenda 2030 with all its seventeen SDG Goals by 2030? This is obviously a *rhetorical* question. It is important such realities are understood and taken into account in seeking appropriate digital inclusion solutions.

## 2.3    Universal Digital Inclusion Challenges for Malawi, Southern Africa

Malawi is a narrow, landlocked stretch of a country in South Eastern Africa covering an area of 118,480 square kilometres, bordered by Mozambique, Zambia and Tanzania. It has a 750-kilometre-long border with Lake Malawi as Figure 2.3 shows.

Malawi is a country of circa 20.4 million people as of 2022 according to the World Bank[98]. The population quadrupled from 4.4 million in 1966 to 18.2 million in 2019. Malawi's population is young, with 46% aged below 15 and 75% below 30 years[99] – which explains why the population is growing so fast.

---

[97] Source: PNG NICTA CEO Keynote Presentation, Digital Transformation Summit 2024, 2nd October 2024, Port Moresby, PNG -
https://www.ict.gov.pg/digitaltransformationsummit2024/
[98] https://data.worldbank.org/indicator/SP.POP.TOTL?locations=MW
[99] Malawi Growth and Development Strategy III - https://npc.mw/wp-content/uploads/2020/07/MGDS_III.pdf

Figure 2.3 – Map of Landlocked Malawi (Adapted from Shutterstock[100])

Like PNG of the previous section, here are some further current challenging specificities about Malawi, because – as I note earlier with the case of PNG – these specificities matter to realising the universal digital inclusion challenge for Malawi too.

i. *Malawi has extremely low Internet and digital inclusion penetration*: as of July 2023, a staggering 80% of the population of Malawi still lacked access to digital technology and services[101], making Malawi one of the least digitally included countries in the world. The Internet penetration rate in Malawi in July 2023 stood at a meagre 20% of the total population, with

---

[100] https://www.shutterstock.com/image-vector/malawi-road-map-154588634
[101] https://www.undp.org/malawi/press-releases/digital-transformation-sustainable-development-undp-validates-initiative-promote-inclusive-digital-transformation-malawi

only 4.03 million Internet users out of a population of 19.91 million (as of July 2023). Figure 2.4 corroborates this latter data by showing a broadband SIM per 100 inhabitants at 39.3%. However, since most urban consumers in Malawi hold 2 SIM cards from both duopoly mobile providers[102], Airtel and TNM, the 39.3% needs to be halved to yielding 19.65% unique broadband subscribers per 100 Malawians – i.e., close to the 20% Internet penetration rate noted earlier.

| ![] Malawi | 118.480 | 20,405,317 | | 172.23 | | 53.33 | | WDI 2023 (2022 data) |
|---|---|---|---|---|---|---|---|---|
| | | | | Nominal | | Monthly GNI per capita | | |
| **Affordability** | | | USD | Africa Rank | % | Africa Rank | | Source |
| | 300 MB prepaid monthly use | | 1.38 | 21 | 2.6% | 39 | | RISS 2023 Q2 |
| | 20 GB prepaid monthly use | | 8.5 | 32 | 16% | 32 | | RISS 2023 Q2 |
| **Adoption** | | | | | | Africa Rank | | Source |
| | SIM per 100 inhabitants | | | | 58.5% | 37 | | ITU June 2022 |
| | Broadband SIM per 100 inhabitants | | | | 39.9% | 29 | | ITU June 2022 |
| **Infrastructure** | | | | | | Africa Rank | | Source |
| | National backhaul km per 10,000 inhabitants | | | | 0.94% | 29 | | NSRC 2023 |
| | Fiber km per 1,000 sqkm | | | | 16.26% | 12 | | NSRC 2023 |
| | IXPs | | | | 1% | 33 | | PCH 2023 |
| | 3G Coverage | | | | 84.4% | 30 | | ITU June 2022 |
| | 4G Coverage | | | | 68.6% | 19 | | ITU June 2022 |
| | Average mobile download speed (Mbps) | | | | 8.3% | 29 | | 29 Cable.co.uk Aug 2023 |

Figure 2.4 – Some 2023 Malawi ICT Data[103]

As Figure 2.4 shows, 3G and 4G coverage stood at 84.4% and 68.6%, respectively, in June 2022. In addition, a typical 20GB prepaid monthly use package costs 16% of the average monthly income in Malawi, way above the 2.6% for a 300MB prepaid package. These statistics highlight the urgent need to bridge the digital gap in Malawi.

Why is Malawi so challenged then on digital inclusion matters? The following specificities of Malawi go a long way to explain why.

ii. Malawi is also *extremely rural – a major challenge to digital inclusion*: one core reason for the meagre circa 20% [of the population] *unique* set of

---

[102] According to the *Malawi Telecoms Review* by Cedar Capital, there were approximately 10 million SIM cards in circulation in 2021, covering about 52% of the population. Source: https://cedarcapital.mw/content/uploads/2021/11/Cedar-Capital-Telecoms-Report-3-November-2021.pdf

[103] Source: Research ICT Solutions - https://researchictsolutions.com/ict-evidence-portal-africa/ict_evidence_portal_africa.php

Internet users in Malawi is due to the extremely rural nature of the country. Only 15% of the population lives in urban centres, though Malawi is one of the fastest urbanising countries in the world at a 3.77% growth rate[104]. The Malawi population density is 211 per km$^2$ according to the World Bank[105] compared to a world population density of 60.29 per km$^2$ according to Macrotrends[106]. This makes it a densely populated country in theory.

For context, the population density of Malawi is almost ten times that of PNG of the previous section, and not too far off the population density of the United Kingdom at 275.27 per Km$^2$. Given it is clearly much more efficient and economical to serve and digitally-include the citizens of countries with high population densities, *why is it not the case with Malawi?* A significant reason for this follows next.

iii. *Malawi is a very 'poor' country – a major challenge to digital inclusion*: about 50.7% of the Malawi population live under the poverty line of below $1 a day[107]. Malawi is one of the least developed countries in the World. It has/had a Human Development Index (HDI) of 0.485, ranking 172 out of 189 countries[108]. This makes disposable income for Malawians to access ICT services and products severely limited amongst most Malawians. While the poverty rate had declined from 52.4% in 2005 to 50.7% in 2015, *the rate of ultra-poor increased from 22.4% to 25% over the same period*[109]. Furthermore, income distribution has worsened over time[110].

iv. Malawi has *many miscellaneous other challenges mitigating against digital inclusion*: these are many other reasons – or "weaknesses" – self-identified by the Malawi Government itself in the Malawian Broadband Plan 2019-2023[111]. They include the following macroeconomic demand and supply-side challenges.

---

[104] Malawi Growth and Development Strategy III - https://npc.mw/wp-content/uploads/2020/07/MGDS_III.pdf

[105] https://data.worldbank.org/indicator/EN.POP.DNST?locations=MW

[106] https://www.macrotrends.net/global-metrics/countries/wld/world/population-density

[107] United Nations Educational, Scientific and Cultural Organisation (UNESCO) Report (2022), https://unesdoc.unesco.org/ark:/48223/pf0000383964

[108] *Ibid.*

[109] *Ibid.*

[110] *Ibid.*

[111] National-Broadband-Strategy-2019-to-2023.pdf (pppc.mw)

- Low GDP and disposable income: as covered above.
- Large, low density, rural population: as covered earlier.
- High cost of terminals and smartphones, and their low penetration.
- Lack of electricity and low dependability of the power grid where there is access.
- High cost of international and national bandwidth despite introduction of third fibre backbone provider: as I introduced, Malawi is a landlocked country whose international fibre connectivity must go through neighbouring countries.
- Lack of local Malawian local language content.
- Shortage of ICT skills and training.
- High taxation levels.
- Low literacy rates.
- Inadequate infrastructure backbone and access within Malawi.
- Lack of competition – only 2 true mobile operators in Malawi, explaining the dual SIM ownership observation.
- Lack of adequate fixed access network, and vandalism of copper infrastructure.
- High cost of capital.
- Quality of Service challenges by network operators.
- Poor business opportunity for Fixed Wireless Access operators (FWA).
- Lack of broadband satellite usage.
- Spectrum release roadmap to further broadband is not yet available.
- Lack of sufficient investment in the sector.
- Lack of implementation and monitoring of existing strategies and plans.
- Telecentres in rural areas are not cost effective or self-supporting.

I find this *self-diagnosed list* of "weaknesses" honest, very comprehensive and very representative of the challenges faced in most Sub-Saharan African countries.

Another *rhetorical* question follows. Does the reader still believe now that Malawi will attain UN Digital Agenda 2030 with all its seventeen SDG Goals by 2030?

Again, it is important such realities are understood and considered in seeking appropriate solutions. It is no surprise then that in July 2023, the United Nations Development Programme (UNDP), in partnership with the Government of Malawi, "unveiled a groundbreaking initiative focused on fostering inclusive digital transformation in Malawi"[112]. This initiative recognizes the vital role of digitalisation in driving progress towards the Sustainable Development Goals (SDGs) in Malawi. The initiative emphasises the need for "purposeful inclusivity, thoughtful design, and human-centric implementation to ensure that the benefits of digital transformation reach all segments of society"[113]. I certainly am unsure about what 'purposeful inclusivity' truly means, but I certainly like the concept of 'thoughtful design' when it comes to digital inclusion.

## 2.4   Universal Digital Inclusion Challenge Across LDCs, LLDCs and Africa in Particular

Of the 2.6 billion long tail still offline, it would come as no surprise to the reader to learn that most of them reside in Africa, and in Sub-Saharan Africa in particular. Figure 2.5 illustrates this clearly using 2023 ITU data.

---

[112] https://www.undp.org/malawi/press-releases/digital-transformation-sustainable-development-undp-validates-initiative-promote-inclusive-digital-transformation-malawi
[113] *Ibid.*

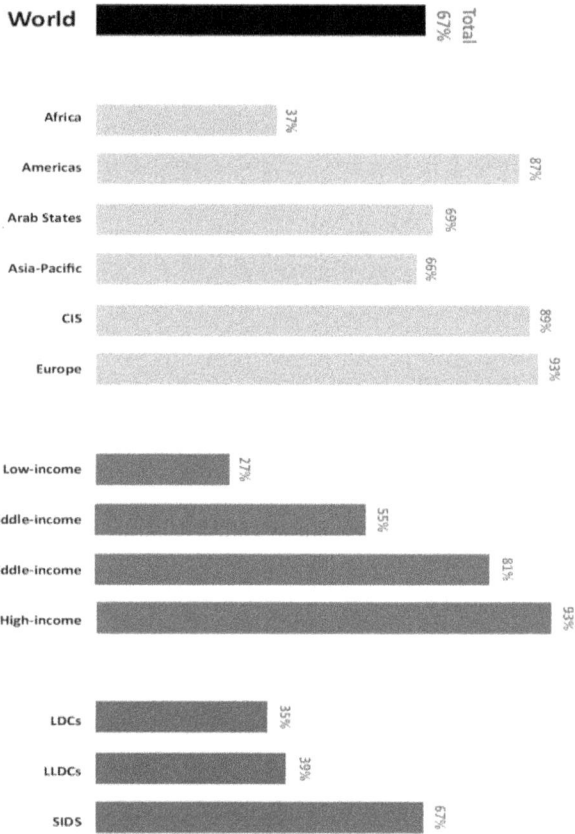

Figure 2.5 – Percentage of individuals using the Internet by region, 2023.
Source: Adapted from ITU[114]

It shows clearly that circa two-thirds of the population of Arab States and the Asia-Pacific regions (69% and 66%, respectively) use the Internet, in line with the global average at 67%. However, the average for Africa (as of 2023) is just 37% of the population. The most recent data for in a report dated April 2025 shows Africa at 38%[115] – the lowest among all ITU regions. The data shows that Internet use remains strongly correlated to the level of a country's development.

---

[114] https://www.itu.int/itu-d/reports/statistics/2023/10/10/ff23-internet-use/
[115] https://www.itu.int/dms_pub/itu-d/opb/ind/d-ind-sddt_afr-2025-pdf-e.pdf

As shown, low-income[116] countries such as PNG and Malawi covered in the last two sections are typically only 27% connected online – as we note in the last two sections, the realities are lower. In contrast, in 2023, 93 per cent of the people in high income countries used the Internet, i.e. they are getting closer to universal digital inclusion – and are likely to meet UN Agenda 2030.

So, universal connectivity is a distant dream in least developed countries (LDCs) as well as in landlocked developing countries (LLDCs) where only 35% and 39% of their citizens are digitally included, respectively (see Figure 2.5).

67% of the population of Small Island Developing States[117] (SIDS) are online – exactly at the world average. For the purposes of this section, Africa is clearly overrepresented in the peoples of the world who are *not* digitally included, as Figure 2.5 shows. Core to this is that Africa is home to many low-income countries as well as many LDCs.

However, following on from the Malawian digital inclusion challenges of the previous section – which applies to many [sub-Saharan] African countries – I think it is important to revisit another key dimension of digital inclusion challenge for Africa. It is one that I believe is often missed in the digital inclusion target-setting board decision rooms of New York, Geneva, London, Hong Kong, Sydney, Washington, D.C, etc. – Africa's size, as illustrated in Figure 2.6.

It still surprises me how many smart non-Africans, including many smart policy makers from Europe, the Middle East, Asia, the Caribbean and the Americas (that I have had the honour of training), have been 'miseducated' about the size of Africa using many current world maps in circulation. They are usually perplexed when I introduce them to Figure 2.6 – and those perplexed have

---

[116] The World Bank assigns the world's economies to four income groups—low, lower-middle, upper-middle, and high income. The classifications are updated each year on July 1 and are based on the GNI per capita of the previous year. GNI measures are expressed in United States dollars (USD). For example, a country with a GNI of less USD $1,085 on July 1, 2022, is a low-income country for the year 2023. A lower-middle income country for 2023 falls with USD $1,086 – $4,255.
https://blogs.worldbank.org/opendata/new-world-bank-country-classifications-income-level-2022-2023
[117] https://www.un.org/ohrlls/content/list-sids

typically included some senior people at organisations like the UN, the World Bank and other august organisations I mention in Chapter 1 (Section 1.5).

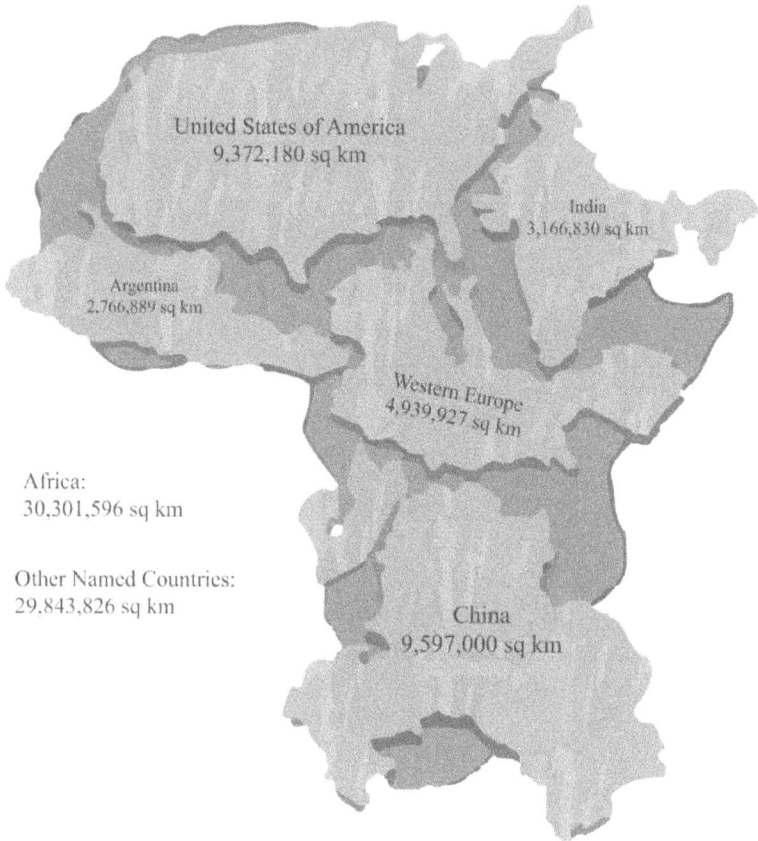

Figure 2.6 - Africa Map: Africa is a Massive Continent
(The USA, India, Western Europe, and China Can Largely Fit into Africa's Land Mass) -
Adapted from Bill Hearmon[118]

So, the reader should keep in mind these <u>three</u> significant further challenges about realising digital inclusion in Africa drawing from Nwana (2014), which are not much acknowledged in many reputable digital inclusion reports out there.

---

[118] William Hearmon, then-chairman at African Broadband Forum, presentation at CTO Forum, Abuja, October 2012.

- *Challenge 1 — the incredible size of the African continent:* Africa is an incredibly big continent, as shown in Figure 2.6. Marvel at the fact that on the African continent you can fit the USA, India, Western Europe, China, and Argentina—and there is still some more land to spare. Africa is truly a vast continent. This means rolling out terrestrial telecommunications networks of any kind is an exercise of humongous costs.

- *Challenge 2 — Africa's 1.5 billion people are more evenly distributed across the continent compared to other continents*: following on from the first challenge is the challenge of the size of the African population at 1.482 billion[119] people (in 2023) and growing, with the sub-Saharan Africa population at some 1.21 billion[120] in 2022. In comparison to Africa, the much smaller populations of big-area countries like Canada, Australia, or the United States of America are mostly concentrated around their coasts.

  o According to the Australian Bureau of Statistics (ABS) in 2001[121], more than eight in ten Australians (85 percent) lived within 50 km of the coast, but by 2019, that proportion had risen to 87%[122]. This equates to over 22 million Australians now calling the coastal areas home.
  o Similarly, 90% of Canadians live within 100 miles (160 km) of the U.S. border. More than 60% of Canadians live south of Seattle[123].
  o Americans also largely live around their three major coasts. In 2000, almost two in three (64 percent) of Americans lived in states along these coasts: 38 percent along the Atlantic Ocean, 16 percent along the Pacific Ocean, and 12 percent along the Gulf of Mexico. It is easier to get at these 'bunched up' coastal citizens.

For these *developed* countries (which are all much smaller in geographical size than Africa – see again Figure 2.6), they can focus their terrestrial

---

[119] https://www.worldometers.info/world-population/africa-population/
[120] https://www.statista.com/statistics/805605/total-population-sub-saharan-africa
[121] Source: www.abs.gov.au - ABS (Australian Bureau of Statistics) (2020b). Regional population, 2018–19, ABS, Canberra.
[122] *Ibid.*
[123] https://bigthink.com/strange-maps/canadians-south-seattle-mental-map-surprise/

telecommunications networks more easily and efficiently where the bulk of their populations live. Africa's fifty-four poorer *developing* countries distribute 1.5 billion peoples more thinly across the continent. To be fair, there is rapid urbanisation in Africa happening too.

Furthermore, *sixty (60) percent of Africans reside in the rural parts of Africa:* this is particularly the case for sub-Saharan Africa according to the World Bank[124]. This – as we have seen in earlier sections – is an incredible challenge for digital inclusion. There is also the low GDP/capita reflecting poverty levels of most of these African countries limiting access to education (hence literacy), education and much else.

- *Challenge 3 — low average revenue per user (ARPUs):* the current 2022 ARPUs on the African continent is in the range of 1.4 to 7.2 US dollars[125] using MTN's data. This is only between 10 and 15 percent of developed countries' ARPUs, yet these revenues must pay for more extensive networks on the *vast* African continent.

The first two challenges (i.e. Challenges 1 and 2) — along with the third, i.e., much smaller ARPUs on the continent — make rolling out financially sustainable and affordable networks for most Africans almost impossible. These overarching challenges must be borne in mind for the rest of the book, as they provide real and extremely costly hurdles to achieving digital inclusion – and the UN Agenda 2030 along with its seventeen SDG goals.

## 2.5   The Exponential Usage Gaps Challenge

Coverage gaps are one thing, but usage gaps are the much harder aspect to the digital inclusion challenge. Figure 2.1 mentions and shows the concepts of 'coverage' and 'usage' gaps which I have glossed over so far.

Figure 2.7 from the GSMA's 2023 Mobile Internet Connectivity Report shows the evidence of the coverage and usage gaps across regions of the world.

---

[124] https://data.worldbank.org/indicator/SP.RUR.TOTL.ZS?locations=ZG
[125] https://www.statista.com/statistics/1076038/mtn-group-mobile-arpu-by-country/

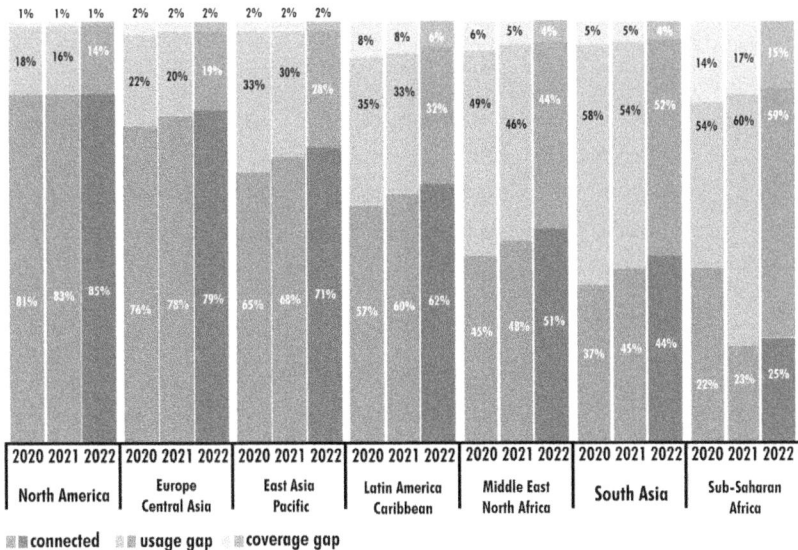

Figure 2.7 - Mobile internet connectivity by region, 2020–2022
Source - GSMA[126]

Permit me to briefly define these two key terms.

There are a couple of ways that lead to people being 'unconnected' to broadband or to the Internet:

i.    Coverage gaps: first, citizens inhabit an area not covered by broadband [typically mobile, or fixed].

ii.   Usage gaps: secondly, citizens inhabit an area that is covered but they do *not* use [mobile] Internet.

The first reason is referred to as the *coverage gap*, because it refers to those peoples who live in an area *not* covered by a broadband network. The second is referred to as the *usage gap*, because it refers to those who live within the footprint coverage of a [mobile] broadband network, but do *not* use mobile internet services.

---

[126] https://www.gsma.com/r/wp-content/uploads/2023/10/The-State-of-Mobile-Internet-Connectivity-Report-2023.pdf

The first is a *supply-side* market failure – i.e., the market is failing to supply some set of citizens, e.g. geographically rural citizens. The second is a *demand-side* failure, the suppliers have provided broadband coverage – yet a certain segment of citizens still do not use it. As we have seen in this chapter, such demand-side failures typically happens because citizens cannot afford to pay for the broadband, cannot afford smartphones to access the broadband Internet, or they just cannot read in the first place. This means *demand-side failures* typically come down to affordability or literacy, or both.

This explains why, according to both Figures 2.1 and 2.7, Sub-Saharan Africa remains the region with the largest coverage and usage gaps. As shown in Figure 2.1, there was some 15% coverage gap (or 180 million people) and a whopping 59% usage gap (or 680 million people) in Sub-Saharan Africa (SSA) by the end of 2022. As Figure 2.7 illustrates, the usage gap in SSA has stubbornly remained at circa 59% or 60% for the years 2020, 2021 and 2022. *It is inescapable that Sub-Saharan Africa's dismal usage gap shows the results of so many supply-side interventions compared to the relative paucity of demand side work.*

There is variation within the SSA region in that "the coverage gap remains much higher in Central Africa (36%) than in Western, Eastern and Southern Africa (where it ranges from 11% to 14%). Meanwhile, mobile Internet adoption is higher in Southern Africa (33%) than in other sub-regions (ranging from 17% in Central Africa to 27% in Western Africa)"[127].

My key message of this section is that the UN's Agenda 2030 (and its SDGs) faces another *massive (and exponential)* hurdle that I do not cover in Chapter 1 in any detail (in certain regions like the SSA) – i.e., the usage gap challenge due to affordability and literacy. How on earth did the UN believe that the SSA could ever eradicate the usage gap challenge by 2030? Coverage gaps are one thing, but usage gaps are the much harder aspect to the digital inclusion challenge. This is one of the true realities of the universal digital inclusion challenge that I feel the UN decision makers in New York, Washington and Geneva fail to appreciate.

---

[127] *Ibid.,* p.11

## 2.6 Summary

The key message of this chapter is one of introducing and reinforcing the insights and predictions from the, admittedly non-scientific, Pareto principle (cf. Chapter 1) in order to demonstrate that the UN 2030 digital inclusion Agenda (along with its seventeen SDGs) are patently *unrealistic* to realise – as concerns most LDCs, LLDCs and/or low-income countries – if not all LMICs too.

The examples of realities on the ground in developing countries like PNG and Malawi covered in this chapter should *disabuse* the inclusion target setters and the reader of the SDGs being realised by 2030. This chapter also deep dives into the general challenges of digital inclusion for LDCs, LLDCs and for Africa in particular, since it is the continent that is least digitally included. It explains why addressing the *coverage gap* problem in Africa is much more difficult and expensive due to the sheer size of the continent, but also how the distributions of the populations in countries like Australia, USA, Canada make the *coverage gap* challenge easier to address in these developed countries.

The chapter concludes by pointing out the reality of the *usage gap* problem in regions like Sub-Saharan Africa, pointing out that this is such a non-trivial problem to tackle – thanks to illiteracy and extreme poverty.

To realise Agenda 2030 and its SDGs even by 2050 – for LDCs, LLDCs and/or LICs – would truly require 'out-of-the-box-thinking' approaches, solutions and leadership.

# 3 Six Essential Traits for Bridging the 2.6 Billion Tail

The long form title of this chapter is *'Six key characteristics required of approaches or solutions to lessening the 2.6 billion long tail connectivity effort'*.

I have argued and 'proven' in Chapter 1 that the well-proven Pareto principle predicts that we have *not* even engaged 86% of the effort – yet – to connect the rather long tail of 2.6 billion digitally-excluded people as at the end of 2023/2024. This is why we desperately need 'out-of-the-box' thinking approaches and [network] solutions to connect these 2.6 billion people. We need this to truly strive to realise universal digital inclusion, meaningful universal connectivity or realising the UN's Sustainable Development Goals (SDGs). Current *incremental* approaches and solutions have patently and visibly failed – and will continue to fail.

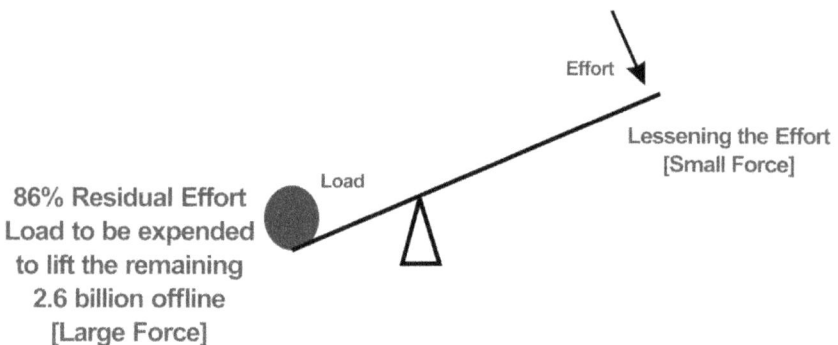

Figure 3.1 – How does a lever work? The 86% 'Large Force' Work to be Levered by 'Smaller' Force Efforts

With Figure 3.1, I illustrate my 'proof' that most of the effort *to be expended* to achieve universal digital inclusion is still ahead of us. However, I argue that many key and relevant stakeholders[128] do *not* realise this 'fact', and that they may already be jaded and tired from connecting 67.5% of the world. 86% of the load effort is still ahead of us, as Chapter 3.1 depicts. So, what are the characteristics (or traits) of the type of 'out-of-the-box' approaches and/or solutions that are required? That is the subject of this chapter. Following are

---

[128] See Section 1.5 in Chapter 1 of the identified relevant stakeholders

some of the characteristics of the sort of approaches or solutions I truly believe are needed.

## 3.1  The 86% Effort Needs Levers and Pulley-type Solutions

One of the five peer reviewers[129] of this book was uncomfortable about the assertion of this subheading – driven by his discomfort with the Pareto metaphor. However, Pareto's rule of thumb holds strongly – arguably buttressed by demand-side usage gap root causes in LDCs. So, I further assert that the 86% load effort ahead of us needs levers and pulley-type solutions. The first characteristic is what I term 'levers and pulleys' characteristic or traits. The realization that 86% of the required effort to digitally connect the remaining 2.6 billion online users is yet to be accomplished may cause some of the prominent organizations referenced in Section 1 (of Chapter 1) to consider withdrawing from these initiatives. I hope they do not.

I argue that what we need, firstly, now are what I consider 'out-of-the-box' approaches, insights and thinking – like the 'levers and pulley' metaphors described next – to connect this unconnected 2.6 billion. Having 86% effort levels ahead of us gives and leaves us no choice, since it has taken a generation to expend just 14% of the effort.

### 3.1.1  Introducing 'Lever' and 'Pulley' Solutions

Permit me to take you the reader (and us) back to secondary school physics where we encountered levers and pulleys. It is a broad consensus that the ancient Egyptians used levers and pulleys in building the incredible feats of the Egyptian pyramids as Britannica attests[130]. The reader who is [still] familiar with these levers and pulley secondary school physics may skip straight to Section 3.1.3.

Starting with a lever – as Figure 3.1 illustrates – a lever is a simple device or machine that works by changing the input [effort] force needed to do work[131] and move some load. The load in this context is the 86% effort load of

---

[129] I completely acknowledge Dr Charley Lewis's discomfort.
[130] https://www.britannica.com/question/How-did-the-Egyptians-build-the-pyramids
[131] Read "to lift the 86% effort 'Load'".

connecting 2.6 billion digitally excluded. We need the effort [on the left of Figure 3.1] to be minimised as much as possible. As Britannica explains, a lever is a:

> "simple machine used to amplify physical force. All early people used the lever in some form, for moving heavy stones or as digging sticks for land cultivation.... A man could lift several times his own weight by pulling down on the long arm. This device is said to have been used in Egypt and India for raising water and lifting soldiers over battlements as early as 1500 BC"[132].

An alternative to the lever is the pulley. A pulley is:

> "a wheel that carries a flexible rope, cord, cable, chain, or belt on its rim. Pulleys are used singly or in combination to transmit energy and motion... Archimedes (3rd century BC) is reported to have used compound pulleys to pull a ship onto dry land. Together with the lever, wedge, wheel and axle, and screw, the pulley is considered one of the five simple machines"[133].

Figure 3.2 shows a 1-pulley system, 3-pulley system and 5-pulley system, demonstrating that more pulleys lead to lesser (and lesser) efforts. Levers and pulleys are typically used with wedges, wheels and axles too.

### 3.1.2 Why Digital Inclusion Solutions Require Lever & Pulley Traits

Why do I propose solutions to the universal digital inclusion challenge need 'lever' and 'pulley' characteristics. Simply, because we need to *lighten* [or make lighter] the **86%** [2.6 billion digital long tail online excluded, and more] effort load. To achieve this requires solutions with the *characteristics* of Figure 3.1's lever system or Figure 3.2's pulley systems. Figure 3.1 illustrates that amplifying or magnifying the 'small force' [on the left] to deliver a much 'larger force' on the right, to lift the 2.6 billion *not* online. Similarly, solutions with *pulley-type characteristics* would also clearly be needed. The 2.6 billion people excluded online, who bear 86% of the effort, need solutions that lower input effort — like the 3- or 5-pulley systems shown in Figure 3.2.

---

[132] https://www.britannica.com/technology/lever
[133] https://www.britannica.com/technology/pulley

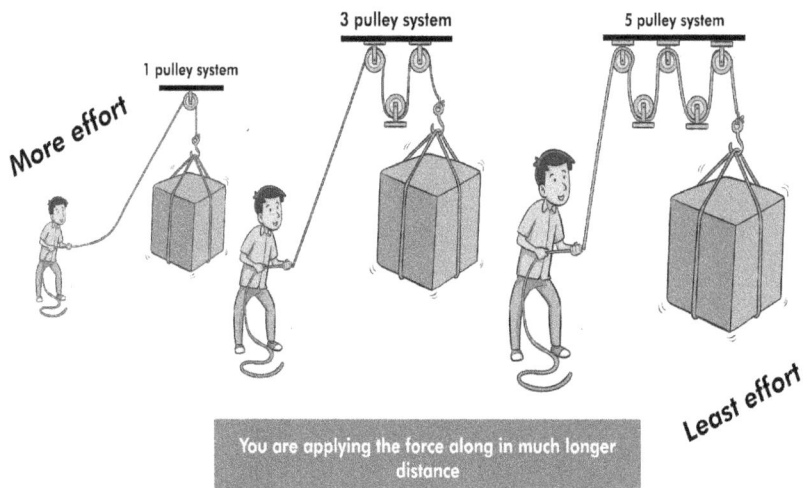

Figure 3.2 – Pulley System Characteristics (Adapted from Source [134])

### 3.1.3 Lever & Pulley Solutions: Speeding Digital Inclusion Efforts

Recall that 86% — or 2.6 billion — represent the online-excluded population. To illustrate this point, consider the 21% of citizens in Papua New Guinea (PNG) who currently lack access to basic 2G [voice] mobile coverage, as indicated in Figure 2.1, along with the 24% not served by 4G [broadband]. Mobile operators — whether acting voluntarily or in response to regulatory mandates — would reasonably prioritise extending service to the most commercially viable 1% segments of the uncovered populations for both 2G and 4G, before progressing incrementally to subsequent portions. Hence, consider these non-exhaustive ten 'sub-loads' for PNG:

i. The 'load' of *increasingly* much higher (i) infrastructure costs and (ii) increasingly costly ownership model to cover each further 1% of the 21% uncovered citizens for 2G voice. This is because these uncovered citizens are increasingly more *rural*, in one of the most rural countries on earth at 87%. Each new [2G/4G] base station covers a lesser number of citizens.

---

[134] https://images.app.goo.gl/6E6v8umvuxnELjwRA

ii. The 'load' of *increasing* costs of power [diesel/kerosene or solar or grid electricity] for powering base stations for each further 1% of the 21%/24% being covered – to provide the telecoms services – and for citizens to power their mobile 2G feature phones and 4G smartphones.

iii. The 'load' of the *increasing* backhauling costs from base stations (e.g. by fibre, microwave and/or satellite) with each further 1% of the 21%/24% being covered. This is due to the inadequate digital infrastructure backbone networks within PNG.

iv. The 'load' of *increasing* unaffordability of even basic 2G feature mobile phones, yet alone 4G smart phones, with each further 1% of the 21%/24% being covered. This is due to the poverty levels for citizens of PNG.

v. The 'load' of *increasing or high* spectrum costs. Mobile operators would need low frequency (sub-1GHz) spectrum bands which propagate further. Such low frequency coverage spectrum tends to be more expensive to acquire from regulators than higher frequency (capacity) spectrum.

vi. The 'load' of *increasing* shortage of ICT/telecom skills and training with each further 1% of the 21%/24% being covered – to train citizens to use their feature phones, smartphones and other ICTs. Telecentres in rural areas are not cost effective or self-supporting.

vii. The 'load' of *increasingly* much poorer citizens being covered with each further 1% of the 21%/24% of uncovered PNG citizens – citizens who may not be able to afford the services anyway. As of 2018, Australian Aid found that "approximately 39% of Papua New Guineans are considered to lie below the international extreme poverty line of $1.90, while 65% lie below the relative poverty line of $3.10. More than 85% of the population live in rural areas, where poverty levels reach approximately 94%"[135].

---

[135] https://marketdevelopmentfacility.org/wp-content/uploads/2019/09/Household-Level-Analysis-of-Poverty-and-Gender-Dynamics-in-Papua-New-Guinea-Final.pdf

viii. The 'load' of *increasing* multidimensional poverty is real: as of 2021, 56.6% of the population of PNG was considered multidimensionally poor, i.e., they experience multiple overlapping deprivations in health, education, and standard of living. In addition, 25.3% of the population is/was classified as vulnerable to multidimensional poverty[136].

ix. The 'load' of *increasing* illiteracy with each further 1% of the 21%/24% being covered.

x. The 'load' of *increasing* cost of money (WACC) and poor business case with each further 1% of the 21%/24% being covered.

I could go on.

Table 3.1 illustrates how the 'levers' and 'pulleys' solutions lessen the 86% 2.6 billion Effort load. Each of the 'levers' or 'pulleys' contributes to lightening load effort – hence accelerating the time to realise the digital inclusion outcomes.

| | Types of the 86% 2.6 Billion Effort | Examples of 'Levers' and 'Pulleys' to lighten the loads and shorten the time to realise the digital inclusion Outcomes | |
|---|---|---|---|
| 1 | The 'load' of *increasingly* (i) much higher infrastructure costs and (ii) increasingly costlier ownership model to cover each further 1% of the | (i) | Lever/Pulley 1 - Much lower 'quality' telecoms infrastructure, and hence much lower costs of Infrastructure (see Table 3.2) |
| | | (ii) | Lever/Pulley 2 - Much lower cost of ownership model costs (see Table 3.2) |
| | | (iii) | Lever/Pulley 3 – Use of Satellite-type innovative solutions that employ similar low-cost 2G feature phones or low-cost 4G smartphones – e.g. evolving HIBS[137] |

---

[136] https://hdr.undp.org/sites/default/files/Country-Profiles/MPI/PNG.pdf
[137] HIBS = High Altitude Platform Stations as IMT Base Stations. HIBS operate in the stratosphere, usually at an altitude of about 20 km. When compared to a terrestrial 2G/4G network, a HIBS system may provide wider coverage. A HIBS system may provide lower

| | | | |
|---|---|---|---|
| | poorer (and poorer) X% uncovered citizens. | | or Direct to Device (D2D) satellite solutions |
| | | (iv) | Lever/Pulley 4 – Community Local Ownership Model – rather than by a major national MNO (e.g. See Figure 3.3 – network run by local Primary School) |
| 2 | The 'load' of *increasing* costs to power [diesel/kerosene or solar or grid electricity] for each further 1% of the poorer (and poorer) X% uncovered citizens. | (i) | Lever/Pulley 1 – Use Solar Power– instead of Grid Electricity or Diesel (see Table 3.2) |
| | | (ii) | Lever/Pulley 2 – Use Locally-supplied 'Hub' power (see Table 3.2) |
| 3 | The 'load' of the *increasing* backhauling costs from base stations (e.g. by fibre, microwave and/or satellite) for each further 1% of the lower (and lower) ARPU X% uncovered citizens. | (i) | Lever/Pulley 1 – Use Consumer VSAT (see Table 3.2) or Starlink-type LEO satellite solutions |
| | | (ii) | Lever/Pulley 2 – Use Locally-supplied 'Hub' backhauling |
| 4 | The 'load' of *increasing* unaffordability of even basic 2G feature mobile phones, yet alone 4G smart phones, for each further 1% of the poorer X% uncovered citizens. | (i) | Lever/Pulley 1 – Government may choose to procure and subsidise 2G feature phones at almost close to free – this way, at least voice services become a designated Universal Service for all citizens. Who pays? This is a country-by-country decision. |
| 5 | The 'load' of *increasing or high* spectrum costs. | (i) | Lever/Pulley 1 – Government and/or Regulator may choose to assign the necessary radio frequency spectrums to Low-Cost Rural Operators (e.g. Figure 3.3) 'for free' |

Table 3.1 – Illustration of how the 'Levers' and 'Pulleys'-Type Solutions Lessen the 86% 2.6 Billion Load Effort

---

latency compared to satellite – and promise to use the same low-cost 2G/4G phones used by terrestrial 2G/4G networks. Thus, in addition to satellite systems, HIBS can play a role for expanding mobile coverage to remote communities.

| Equipment | Traditional Typical Cost | Local Cost |
|---|---|---|
| GSM/2G/4G Access Point | Medium to High | Low (Endaga) |
| Mount | High (Tower) | Very Low or None (Roof or Tree) |
| Backhaul | High (Microwave/Fibre) | Low (Consumer VSAT) |
| Power | High (Diesel) | Low (Local or Solar) |
| Fence | High (Protect diesel and more) | None (on Owner's property) |
| Road/Transport | High (bad Roads) | None (Local Transport) |
| Total | USD $150,000 – $250,000 | USD $10,000 – $15,000 |

Table 3.2 – Lower Infrastructure Costs & Lower Ownership Model. (Original Source: Endaga (2014)[138] – adapted with permission of the Original Owner, Lance Condray)

- Location:
  - 4 hours drive from nearest traditional cellular coverage

- Network
  - Live in February 2013
  - 400+ Subscribers
  - **Run by local primary school**
  - Interconnects to major national carrier

Figure 3.3 – A Low-Cost Ownership Model Deployment on a Tree run by a *Local Primary School* in Papua, Indonesia (Original Source: Endaga[139], adapted with permission of the Original Owner, Lance Condray)

---

[138] Endaga was acquired by and integrated into Facebook (today Meta) in 2015.
[139] Endaga was acquired by and integrated into Facebook (today Meta) in 2015.

## 3.2 Using Hubs-and-Spokes to Lessen the 86% Load Effort

As outlined previously, the first characteristic essential for reducing the 86% — or 2.6 billion — load effort is the concept I have referred to as the 'levers and pulleys' approach. The second necessary element involves the 'hubs and spokes' structure, which must also be incorporated into solutions aiming to decrease the significant load effort ahead.

### 3.2.1 Introducing 'Hubs and Spokes' Solutions

The *hubs and spokes model*[140] is a network design where a Central Hub (or coordinating organisation) connects to multiple edge or peripheral nodes (or spokes), without the spokes directly connecting to each other. The spokes represent partner organisations that are linked to the hub. This model – depicted in Figure 3.4 – is commonly used in various network contexts beyond ICT infrastructure and telecommunications, including transportation, railways, banking and more.

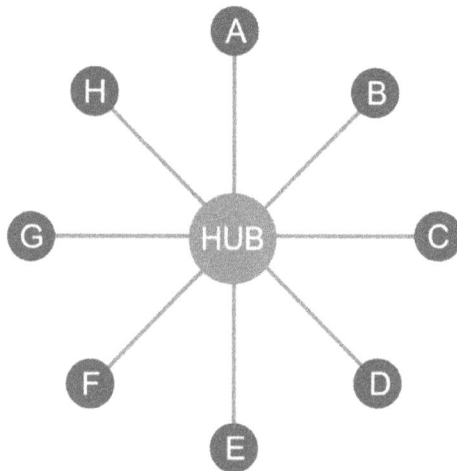

Figure 3.4 – A Hub and Spokes Model
(The Hub could be in the cloud)

---

[140] A mesh model version variation of this will offer better redundancy.

Using Figure 3.4, the central Hub – some aspect of which could also be in the Cloud – could be coordinating in the following example contexts:

(i)     Coordinating the efficient supply and management of *power/electricity* to all the schools, health centres, churches, etc. – represented by Spokes A to H, such that the Hub efficiently manages the loads across all of the spokes.

(ii)    Coordinating the efficient supply and management of *fibre, TVWS or satellite bandwidth* to the Spokes A to H, such that the procurement of expensive fibre or satellite capacity is done once only by the Hub organisation.

(iii)   Coordinating the efficient supply and management of ICT [Cloud-based] services, wherein the main servers and server rooms are housed in the Hub whilst the much 'simpler' client computers are used at the 'spokes' organisations.

(iv)    Etc.

Therefore, there are some noticeably clear efficiency benefits of this model:

(i)     *Centralised Hub Coordination*: the hub ensures that all spokes are working synergistically towards common goals, minimizing duplication of efforts and promoting alignment of strategies.

(ii)    *Cost-Effectiveness*: by centralizing operations, redundancy at the spokes is reduced, and operational costs are reduced to a minimum. Consider operations like security of fuel supplies at all spokes, which is clearly more expensive than securing all key such fuel assets more efficiently at a single Hub.

(iii)   *Efficient Communication:* the direct lines between the hub and each spoke ensure clear and consistent communications, reducing potential communication mishaps.

(iv)    *Streamlined Resource Allocation*: the coordinating hub has an overseeing view of the entire network, allowing for effective and efficient resource allocation based on strategic needs.

(v)     *Scalability*: the hub model can easily scale to accommodate growth and expansion, with the addition of other spokes whilst the capacity at the coordinating Hub is also increased.

(vi)    *Improved Connectivity*: enhancing the communication and coordination between the hub and spokes improves overall network performance.

The hubs-and-spokes model may encounter challenges beyond the redundancy constraints I note earlier. This is because it introduces single points of failure. For instance, certain 'spokes' may need to be converted into 'hubs', resulting in traffic routing inefficiencies. Therefore, it may not always be the most efficient model, but the underlying logical reason for using such a model remains.


### 3.2.2 Hubs-and-Spokes Solutions: Accelerating the 86% Effort

So, how do solutions with 'hubs and spokes' characteristics lighten the 86% 2.6 billion Effort – shortening the time to outcomes? I have been advocating for such a hubs and spokes model for more than a decade now, ever since I observed (and still observe) Universal Service Funds[141] (USFs) using their much limited funds to install desktop computers, Internet connectivity (typically via satellite) and other ICT equipment in rural schools and health centres. These 'fat' power-hungry computers and onsite 'server rooms' hosting servers to serve these desktop computers naturally require much power that such rural contexts just do not possess. The fate of these 'investments' are usually predictable. The school and health centre authorities are unable to pay for the high costs of maintaining these ICTs because other priorities precede ICT, and – inevitably – all of the computers/ICT fall into disrepair and disuse – and are usually stolen anyway. Therefore, I usually advocate for a hubs and spokes model for ICT Connectivity in rural areas.

To explain the benefits of hubs and spokes model, consider the following typical *contexts* in countries like Malawi or PNG.

1.  **Typical Rural/Semi Rural Primary Schools Contexts**: rural primary schools in a mid-scale Sub-Saharan Africa or South East Asia country, like Malawi or PNG, respectively, would typically consist of (most LDC citizens will recognise these):

---

[141] A Universal Service Fund (USF) is a system of telecommunications subsidies and fees managed by Governments and/or Telecoms Regulators Its primary goal is to promote universal access to telecommunications services, particularly across rural and unprofitable areas.

- *Classrooms*: a typical primary school in such contexts as rural Malawi or PNG has circa 5 to 10 classrooms with basic structures for teaching, often with simple furniture like desks and chairs – and sometimes with hardly any secure doors.
- *Teachers/Teachers Office*: perhaps five to a dozen teachers including the head teacher, and a small teachers' office for teachers to prepare lessons and store materials.
- *Clean Water*: access to clean drinking water would typically be a challenge in some or many rural areas.
- *Toilets and Sanitation*: toilets are basic pit toilets and typically inadequate with poor hygiene standards, with hardly any handwashing stations.
- *Play Area*: there would typically be a reasonably large outdoor play area for pupils to engage in physical activities.
- *Electricity, Computers and Internet Connectivity*: these schools typical face hurdles such as limited access to electricity, computers, and Internet connectivity. Many would not, or hardly, have any such facilities.
- *Physical Security*: what security? Security is poor at such schools and they hardly have secure rooms to 'house' server rooms and electricity grid connections to such schools are typically non-existent.

2. **Typical Rural/Semi Rural Health Centre Contexts:** rural primary health centres in a poor Sub-Saharan Africa or Asias country like Malawi or PNG respectively typically consists of (most LDC citizens will recognise these):

- *Outpatient Area/Room*: an area for patients who need medical consultations and minor treatments without staying overnight.
- *Inpatient Wards*: perhaps a couple of inpatient wards for patients who would need overnight stays for treatment and monitoring. Hygiene in such wards is usually a challenge with patients comingling with their family members helping them out.
- *Emergency Services Room*: there may be an emergency area with some basic equipment for urgent medical cases.
- *Maternity Ward*: there typically would be a dedicated room for delivery services, though prenatal and postnatal care may be a challenge.
- *Toilets and Sanitation*: Toilets are basic pit toilets and typically inadequate with poor hygiene standards, with hardly any handwashing

stations. Some (few) health centres would have flush toilets. Water supply for flush toilets would typically be challenging.

- *Water Supply & Clean Water*: access to clean drinking water would typically be a challenge in some or most rural areas.
- *Laboratory/Pharmacy*: sporadically, these may be provided – but typically, rural patients are sent to do their medical tests in private test centres and purchase their medications from private pharmacies in distant urban areas.
- *Electricity, Computers and Internet Connectivity*: these rural health centres would have poor and interrupted access to electricity – typically powered by generator(s) fuelled by diesel or kerosene. There may be several computers on site, and poor Internet connectivity through a satellite VSAT[142]. Others would typically not have any such facilities.
- *Physical Security*: Like with the rural primary schools context above, physical security would be average at best, and there would be challenges in securing their limited ICT infrastructures and fuel for their electricity generators.

I provide these above depictions of typical rural [semi-rural] primary school and health centre contexts in the context of connecting the 2.6 billion not online for three core reasons:

(i) *Other higher priorities than online connectivity exist within these contexts*: the reader can observe that these contexts clearly lack other basic priorities including clean water, basic toilets and sanitation, clean maternity wards (or clean medical wards general), and basic electricity – let alone computers and Internet connectivity. These other priorities are evidently higher.

(ii) *Physical Security for Valuable Assets*: these contexts do *not* allow for easy physical security. Consider a rural health centre. Securing the diesel or kerosene for their electricity generators is a major headache, with break-ins and theft of such fuel a normal occurrence due to poverty.

(iii) *ICT, Internet & Connectivity Capacities and Skills*: these are hard to find and maintain in these rural contexts. ICT staff at these rural locations

---

[142] VSAT stands for Very Small Aperture Terminal. This is a type of satellite communication system that uses small, portable earth stations to provide Internet, data, and voice connectivity. VSAT systems are widely used for remote locations where terrestrial Internet or communications infrastructure is unavailable or unreliable

have extraordinarily little to do – and typically leave for other lucrative opportunities in urban areas.

For these latter three core reasons and more, a 'hub and spokes' model is a much *more logical and rational network* model to have. The Hub 'houses' all the key 'stealable' assets [computers, servers, kerosene, diesel, other key supplies, etc.] and provides good physical security (see Figure 3.4) 'once' in such a network. Similarly, enough ICT tasks have been 'aggregated' at the Hub concerning the online connectivity of the many 'spoke' locations. Perhaps the Hub could be several kilometres away from the spoke locations.

## 3.3　Ultra-Light ICT Solutions for the 86% Effort at Spokes

My third proposed trait stipulates that the 86% Load Effort requires exceptionally light ICT Solutions at the spokes. In the last section, I narrate how I have been advocating for a hubs and spokes model for lessening the 86% Load Effort in Rural areas ever since I observed Universal Service Funds[143] (USFs) using their much limited funds to install desk top computers, Internet connectivity and other ICT equipment in rural schools and health centres. In it I railed against these 'fat' power-hungry computers and onsite 'server rooms' hosting servers to serve these desktop computers that naturally require much power that such rural contexts just do not possess. The obvious solution to this problem is to have exceptionally light ICT solutions at the 'spokes' locations (see Figure 3.4), e.g. at rural primary schools and/or rural Health Centres.

### 3.3.1　The Need for Ultra-Light ICT at Hub Spokes

Exceptionally light ICT solutions at the spokes of the Hub and Spokes networks offer several important benefits:

---

[143] A Universal Service Fund (USF) is a system of telecommunications subsidies and fees managed by Governments and/or Telecoms Regulators Its primary goal is to promote universal access to telecommunications services, particularly across rural and unprofitable areas.

(i)     *Lower Costs and Cost Efficiency*: lighter ICT solutions often require much less [electrical] power, less investment in hardware and software, licences, maintenance costs, etc. – making them more affordable for rural or remote organisations.

(ii)    *Scalability*: exceptionally light ICT solutions at the spokes [of the hub and spokes network] can be easily scaled up or down based on the needs of the spoke organisations, allowing for flexibility in the allocation of key assets or resource, e.g. fibre/satellite bandwidth, on-grid or off-grid electricity and even human ICT resources.

(iii)   *Ease of Implementation*: very light ICT solutions at the spokes are typically easier to implement and maintain, reducing the need for extensive technical expertise and perhaps even on-site [at the spokes] human ICT experts.

(iv)    *Focusing on Priorities & Resource Optimisation*: by using exceptionally light ICT solutions, rural or remote organisations can optimise their resources, focusing on core priority activities (like clean water, clean wards, clean toilets with handwashing stations, sanitation, etc.) – rather than on managing complex IT systems.

### 3.3.2  Mwabu[144]: a Light ICT Solution for Primary Education

An example very light ICT solution is offered by Mwabu's Primary School Teaching Technology. Mwabu – based in Lusaka, Zambia – provides interactive digital learning solutions for primary schools in sub-Saharan Africa.

It was founded in Zambia in 2010, and Mwabu Zambia has been instrumental in improving educational outcomes by supporting teachers and learners with using ICT technology[145]. As I cover later, its light 'spokes' solution consists of a handheld smartphone and a portable projector (see Figure 3.6).

As Figure 3.5 shows, adopting the strapline 'Where teaching meets technology', Mwabu provides ICT solutions for parents, for teachers, for schools and for organisations. Mwabu's comprehensive curriculum provides digital content for subjects like *English, Maths, and Science* and its content includes interactive lessons, animations, and activities that are culturally relevant.

---

[144] Mwabu - https://www.mwabu.com/
[145] *Ibid.*

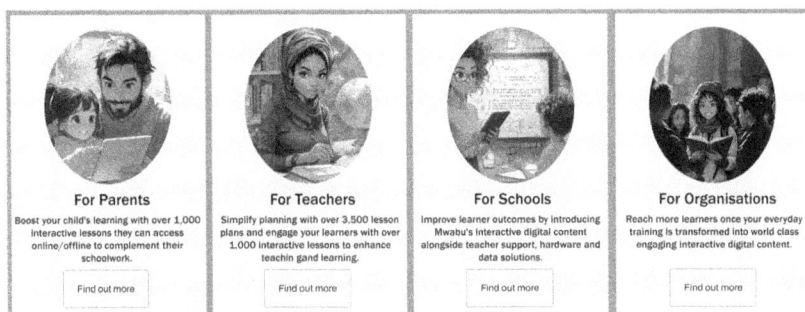

Figure 3.5 – Mwabu's 'Where Teaching meets technology"
Source: Adapted from Mwabu [146]

Mwabu provides teacher training and ongoing support to help educators to effectively use digital tools in their classroom. Mwabu has used its very hard-won experience to build an educational-technology platform that helps all teachers, even those who reside in rural areas.

So, Mwabu has not only developed but been delivering its digital curriculum-aligned content into rural Zambia since 2013. Mwabu utterly understands the challenges faced by teachers and learners, including those in rural areas. They collaborate with various organisations, including Zambian Ministry of Education and UNICEF to deliver digital educational content. Mwabu claims that studies have shown significant improvements in literacy and numeracy among students using Mwabu's resources.

Figure 3.6 depicts the two core elements of Mwabu's Very Light 'spoke' ICT technologies: a handheld smartphone and a projector. Mwabu's entire content bank[147] is downloadable onto a smart phone. This provides a teacher – even an untrained one – with the resources needed to deliver a *quality* lesson. This, allied with the use of battery-powered solar charged display devices (e.g. projectors) and a system for data gathering, raises teaching quality.

---

[146] *Ibid.*

[147] This includes essential educational tools (content, lesson plans, assessments) and teacher training and support tools – which can all be deployed using the most ubiquitous piece of hardware found in rural areas – the mobile phone.

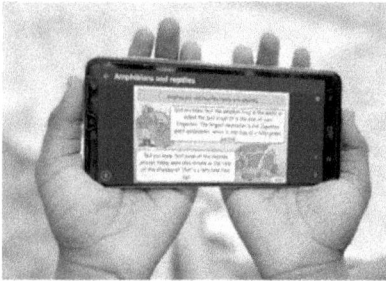

Figure 3.6 – Mwabu's Very Light ICT Technologies
Source: reproduced with permission of the Owner, Ian McFadyen (Owner of Mwabu)

So, there is *no need* to introduce expensive computer desktop hardware, software and their maintenance into all rural primary schools. As the last section explains, this is not feasible, due to the inherent need for power, connectivity and IT support at 'spoke' rural schools.

The entire technology set shown in Figure 3.6 is also ready for *offline use* too, through the Mwabu Learn app. This means the mobile device could be pre-loaded with all the learning information, particularly for primary or secondary education – not necessarily for university education.

Mwabu is unquestionably an innovative approach to developing technical solutions for delivering ed-tech in rural and under-served areas. However, I acknowledge – as one of the peer reviewers of this book commented – "that still leaves a digital divide between Mwabu learners and Lusaka [the capital] learners"[148]. It is truly non-trivial to 'level' the world.

---

[148] Dr Charley Lewis is right to point this out.

## 3.4 Redefining Broadband as a Universal Service for the 86% Effort

For my fourth proposed essential trait to bridging the 2.6 billion digital divide, I posit the 86% Load Effort demands a redefinition and rethinking of broadband connectivity as a 'universal service'. Due to the unique and *personal [ownership] nature* of the mobile phone, the goal of connecting the long tail of the unconnected 2.6 billion has come to be *inadvertently and mistakenly* defined 'as each of the 2.6 billion should be able connect online when they want to and using their own individual connectivity device'. This is largely because this is the case (or model of connectivity) for the 5.4 billion who are online today. However, I make the case in this section that this 'model' or thinking should be relaxed for the unconnected long tail 2.6 billion – many of whom reside in Sub Saharan Africa, Latin America and parts of Southeast Asia at least – a significant percentage of who can neither read nor write. There is a reason why there is a distinction between 'access' and 'service' in any best-practice Universal Access Framework.

So, I argue we should revert to the thinking and definition of 'universal service' – adopted by the telecoms industry – *before* the coming of mobile communications and the ubiquitous mobile phone/device. Many countries have developed and continue to develop Universal Service Obligations (USO) across several basic services like access to clean water, access to food, access to sanitation services, access to maternity services, etc. How can *commodities* like basic drinking water, phone communications, waste management, electricity, basic bank accounts, basic literacy, basic health services, etc., not be available to billions of the 8 billion+ of humanity that live today? Whether these USOs are functional and working in most developing countries is frankly rather moot. My central point of this section is that a 'universal service' is any service that the Government expects to be available, affordable and accessible throughout the population (Nwana, 2024). Therefore, a universal service has two key aspects to it. First, it means the service must be made available to all $\underline{X}$, perhaps within a given area or of the entire population. Second, it should be made available at a uniform and affordable price. To achieve the latter, the Government may or would have to employ subsidies. It is important to point out some key definitions and nuances herein:

- 'Universal Service' defines those services deemed or designated by the Government as "essential" or central for daily modern living to most of its population.
- A 'universal' service must be made *available* to all X, but it is up to Governments to specify what X refers to. As Nwana (2024) explains, the reason is because X would typically represent different descriptors depending on the designated essential services and the countries concerned. X (qualifying as "universal") may be referring to the following descriptors depending on the services designated:

  ✓ "Individual": the notion of universal for a service like basic 2G mobile telephony services or banking services where a Government may want every citizen to possess a basic mobile device with their own individual number along with an individual bank account, respectively. This is the 'intentionality' behind the case with the famous Aadhaar Identity System[149] in India.

  ✓ "Community": the notion of universal for a service like post offices may be delimited to refer to a defined community, e.g. every district "community" must be endowed with at least one post office.

  ✓ "Location-based" or "Specified geographical area-based": the notion of universal for a service like mobile/cellular is strictly speaking location-based (or specified geographical area-based) because telecom regulators know the exact locations of telecoms transmitting base station equipment, and therefore they can predict [with good technical accuracy] the coverage area of the mobile/cellular service, and hence the availability at specific geographical areas.

  ✓ "Homes": the notion of universal of certain services like human wastewater management (covering toilet waste) would typically be delimited to cover homes.

  ✓ "Streets": the notion of universal for a service like home waste collection services would typically be delimited to all streets on which homes are situated.

---

[149] https://en.wikipedia.org/wiki/Aadhaar - As of October 2024, Aadhaar has over 1.383 billion holders, making it the world's largest biometric ID system. However, while Aadhaar is widely used, it is not mandatory for all services. The Supreme Court of India has ruled that no person should suffer for not having an Aadhaar number.

- ✓ "Offices": the notion of universal of a business broadband and Internet service may be delimited to target offices and some selected home offices too.
- ✓ "Entire country": some services like satellite broadcast services can cover the entire country – but at what level of Quality of Service (QoS)?

A key message of this entire section is that the word "universal" itself is *typically service and/or geography-specific* and more – and not necessarily 'individual' as has come to be defined in the telecoms industry – and which arguably drives the effort levels behind the 2.6 billion not online today.

In plain English, the 86% 2.6 billion unconnected load effort can be significantly 'lessened' (and made much lighter) by *policy and regulatory definition*. If the Government and/or Regulator defines X as some 'Community', some 'geographical area' or as 'homes', it simplifies the problem much compared to X referring to an 'Individual'.

If there is a Hub central location that provides broadband Internet services within a kilometre radius of every home, that enable all citizens to have the ability to connect online – this would be much better than nothing and may/would suffice for many village and/or rural contexts in PNG or Malawi.

This is no different from 'universal access' to hospitals, or schools or libraries. No student/patient has their own 'individual' school or hospital.

*Usage Gap = Universal Access (coverage gap) – Universal Service*
- **The Usage Gap Equation**[150].

The 'Usage Gap Equation' clarifies this section's message. Redefining 'Universal Service' to include more citizens, even with a shared definition, significantly reduces the Usage Gap.

---

[150] Peer Reviewer Dr Charley Lewis 'derived' this formula, so I take no credit for it.

## 3.5  Creating Awareness & Empowering Bottom-Up, Community-Led Ownership for the 86% Effort

For my proposed fifth essential trait of solutions to bridge the 2.6 billion digital divide, I posit that the 86% Load Effort also demands creating awareness and empowering a 'bottom-up' 'community-led' ownership mindset. I argue that – in addition to the top-down nature of all the prior four characteristics, a key proportion of the 2.6 billion *not online:*

*(i)*   need to be made aware that they are unconnected (along with its benefits) – if you do not have the Internet, you cannot imagine its benefits; and

*(ii)*  will 'need to take matters into their own hands' and own the problem.

Clearly, both *competition* and the best efforts of *regulation* are failing them for whatever demand-side and supply-side market failure reasons (see Nwana, 2024).

*Empowering citizens and consumers* is always a key goal of good regulators. This is because, ultimately, no one feels the pain of the opportunity costs of not being to go online like the disenfranchised or unconnected community itself – the high costs of diesel, the frequent and expensive trips to the local town to get connected, the lack of online services to enable education and agriculture, etc. Therefore, policymakers and regulators must *enable and empower* the disenfranchised and unconnected communities to a significant degree to take 'ownership' of the problem of them not being online. Some 'enlightened' community members are generally able to make a significant difference. The Community could be enabled through a 'bottom-up' spectrum policy (access to 'free' radio spectrum in rural areas), direct subsidies to community organisations and their leadership, etc.

Rural and disenfranchised communities who are *not* online can take the lead in several ways to address the challenge of limited or no Internet access:

(i)   *Community Networks*: community-owned and operated broadband networks can be established that provide affordable and reliable Internet access. As seen in Figure 3.3, a Local Primary School in Papua, Indonesia owned/owns and operated/operates a low-cost network with a deployment on a tree run by the school. This was driven from within, as opposed to externally conceived. So, community networks are often built and maintained by the community, ensuring that the network infrastructure meets the specific requirements of the community. A key problem with Community Networks is what happens when people leave the communities, highlighting the problem of lack of ubiquitous access. This has been a key issue killing such networks in Nigeria and South Africa.

(ii)    *Lobbying for Preferential Government Initiatives & Engaging Local Leaders:* through communities truly 'owning' their lack of online connectivity, they tend to lobby and advocate Government and local leaders for [Government] policies, programs and regulations that prioritise rural Internet and broadband access. These can lead to increased funding and support for broadband/Internet infrastructure projects. Engaging local leaders and organisations to raise awareness about the importance of Internet access and advocate for solutions can also drive change at the community level. An example of such lobbying is the *#DataMustFall campaign*[151] in South Africa – a powerful example of grassroots lobbying that successfully influenced national policy on digital access.

Initiated in 2016, the campaign was spearheaded by civil society organizations such as amandla.mobi[152], which mobilized communities – particularly low-income Black women – to advocate for reduced mobile data prices. Participants organized protests, submitted public comments, and shared personal testimonies detailing how high data costs exacerbated inequality.

Their endeavours resulted in several significant outcomes:

---

[151] https://mybroadband.co.za/news/telecoms/341887-datamustfall-the-complete-story.html
[152] https://amandla.mobi/data-must-fall-campaign/

- The Independent Communications Authority of South Africa (ICASA) introduced pro-poor regulations, including mandatory data usage notifications and rollover options.
- Major providers like Vodacom and MTN implemented price reductions of 30–50% on certain data bundles, benefiting over 13 million individuals.
- Government acknowledgment, with President Jacob Zuma in 2017 and subsequently President Cyril Ramaphosa in 2019, prioritising data affordability in national addresses.

This campaign exemplifies how community-led advocacy, coupled with strategic engagement with regulators and policymakers, can influence digital inclusion policy. Furthermore, it highlights the importance of narrative power in transforming lived experiences into political leverage.

(iii) *Public-Private Partnerships*: local community ownership [i.e. 'owning' their lack of online connectivity] and leadership is more likely to lead to creative collaborations with private companies, non-for-profit foundations, foreign NGOs, universities and Government agencies in order to help bring Internet/broadband infrastructure to rural areas. These partnerships bring funding, human resources and expertise to build and maintain networks.

(iv) *Trialling Innovative Technologies:* local community ownership may choose to explore alternative technologies to being connected beyond 2G, 3G, 4G and 5G technologies provided by MNOs and the traditional mobile/cellular industry. These could include technologies such as Wi-Fi, TV white spaces (TVWS) technologies for affordable broadband (see Nwana, 2014, Chapter 6), satellite Internet through the likes of StarLink/Oneweb LEOs, fixed wireless access (FWA) technologies, etc. – that can provide Internet access in areas where traditional broadband is too uneconomical and/or not feasible.

(v) *Digital Literacy and ICT Training Programs*: I have seen some motivated local community-led digital trainings improve digital literacy in rural Kenya[153], rural Nigeria and elsewhere – which are far superior to any trainings that could have been dreamed up from the urban areas. This is usually because the trainings are in response to real local needs, e.g. a local Community Women's Cooperative of Entrepreneurs in rural Nigeria sought and received both digital literacy trainings, including on how to use ICT tools to better manage their Cooperative[154].

By rural communities being empowered to own their challenges – including lack of broadband Internet access – some would be able to overcome the challenges of limited Internet access and ensure that all residents can benefit from digital connectivity.

## 3.6 Broadband Stakeholders: Driving Collaboration for the 86% Effort

The 86% Load Effort also demands collaboration and cooperation amongst broadband stakeholders. Collaboration and cooperation amongst the myriads of Broadband Stakeholders worldwide mentioned in Chapter 1 is vital and invaluable to lessening the load of lifting the tail 2.6 billion who are *not* online.

I do not write this lightly, because I see so much duplicative activities ongoing worldwide. I also observe so many *sub-scale* broadband digital inclusion projects which are completely unlikely to 'move the needle' of connecting the long tail 2.6 billion, including ones that I have been involved with, like Civilisation in a Box that I describe in the next chapter (Chapter 4). What if there was some true and meaningful cooperation and collaboration on *big scale[155]* broadband inclusion projects by a coordinated *subset* of all the well-meaning broadband inclusion stakeholders that I elaborate upon in Section 1.5 (Chapter 1)? There are literally dozens of them. I think it would need a strong facilitator too.

---

[153] https://voiceout.or.ke/how-kenyan-youths-are-empowered-through-digital-literacy-training/
[154] https://www.techherfrica.org/tech-herfrica-and-other-csos-expand-digital-literacy-and-e-commerce-to-rural-women-entrepreneurs-in-northern-nigeria/
[155] e.g. entire country or a continental region

I return to this issue of collaboration, cooperation and strategy as a key lever in Chapter 7.

## 3.7  Summary

This chapter has introduced six essential characteristics of the type of approaches and network solutions required to lighten the load of lifting the long tail of 2.6 billion online, the first two being 'levers and pulleys' and the 'hubs and spokes' characteristics. The third is that of having exceptionally light ICT technologies at the spokes organisations which allow for scalability and cost efficiency – allowing for easier lessening of the 86% 2.6 billion offline 'load'. The fourth is that the 86% Load Effort demands a wholesale redefinition and rethinking of 'universal' broadband connectivity. This would significantly lessen the pressure on LDCs and make it consistent with some of the challenges they still have with other issues like health and schooling. The fifth essential characteristic is one of enabling true community ownership of the problem (of lack of online connectivity) through clear policy and regulatory steps. The sixth and last characteristic is that of collaboration and cooperation amongst the scores – if not hundreds – of well-meaning broadband digital inclusion stakeholders – some of which I mention in Chapter 1. I cover this in Chapter 9.

A peer reviewer[156] provided feedback specifically on Section 3.4, and more broadly, on the entire chapter.

> "There is a political challenge inherent in this approach – people don't aspire to 2G handsets, long-drop toilets, barefoot medicine or rural schooling. And there are other plenty of examples of the difficulty implementing such models".

This insight is so profound that I felt it necessary to include it at the conclusion of this chapter. My brief remark in response, if it may even be considered a response, is to recall Voltaire's wise saying[157]: *let us not allow the good to*

---

[156] Thanks to Dr Charley Lewis.
[157] This saying originates from the French phrase *Le mieux est l'ennemi du bien*. This was popularized by Voltaire in his *Questions sur l'Encyclopédie* in 1770 - https://en.wikipedia.org/wiki/Perfect_is_the_enemy_of_good.

*become the enemy of the best.* Beyond this, I completely acknowledge this critique and invite others to improve on my approach.

# Part II: An Innovative Solution and Some Cautionary Warnings

Part I of this book (i.e. Chapters 1, 2 and 3) has established three *summarised core categories* of reasons <u>why</u> the 'Tail 2.6 billion' are *not connected* online as of 2025 (cf. Figure 1.1) – as well as alluded to <u>why</u> a significant fraction of the 'Mid-Tail 2.7 billion' are *not meaningfully connected.*

The summary three categories include demand-side market failures, supply-side market failures and policy/regulatory/leadership failures. It is crucial to keep <u>the details</u> of these three categories in mind as we develop innovative solutions to address them or at least mitigate them.

1. **Demand-side market failures**
   a. *Poverty (general poverty):* general poverty is a key barrier for the poor. I have been to regions in Africa and Asia where people literally do not have roofs over their heads at night. They sleep under trees or make-do tents at night, open to the elements. Internet connectivity to these people is not an option at all. I call this crushing poverty.
   b. *Poverty (low income): affordability* is a key barrier for the poor, including affordability challenges of Internet-enabled handsets.
   c. *Low-priority concern*: connectivity is a lower priority concern compared to other more basic 'essential' services like clean water & sanitation, basic daily food and nutrition, school/education fees for children, healthcare, etc.
   d. *Literacy/education challenges:* how much good is the Internet to the illiterate?
   e. Lack of *Awareness:* I have personally observed lack of awareness of broadband/Internet and its potential benefits bedevil small businesses in my native Cameroon in Central/West Africa.
   f. *Lack of digital skills*: even the literate need to be digital-skilled in order to make the Internet useful to their lives. Lack of digital skills and confidence to use the Internet is a key barrier.
   g. *Low population density (rural areas):* this renders the economics of connecting such regions unviable, leading to no network availability, due to *uneconomic demand.*

h. *Exorbitant usage-based pricing models*: in Part I, I note how connectivity in low-income countries is largely usage-based and quite expensive. This minimises demand, mitigating the realisation of any definition of 'meaningful connectivity'.

i. *Miscellaneous other issues*: some people are being prevented from accessing the Internet due to social norms. This category also includes safety and security concerns in certain regions of the globe, including national security. Fears related to online safety and privacy also deters some people from using Internet services. There are also gender issues that abound.

All of the above issues contribute to the *usage gap* challenge, wherein billions of people live within reach of a [mobile] broadband networks but do *not* use the [mobile] Internet.

2. **Supply-side market failures**
   a. *No electricity/power*: leading to <u>no availability</u> of networks.
   b. *Poor basic infrastructure like roads and rail*: to enable and facilitate telecoms networks to be built and facilitating the logistics to fuel and/or maintain such infrastructures.
   c. *Unviable [infrastructure] business models*: no rational entrepreneurs and/or businesses will fund telecoms networks characterised by business models that would not be viable.
   d. *Lack of suitable 'hubs and spokes' ICT solutions*: low-income economies need more ICT solutions with this *architectural* characteristic (and others I list in Chapter 3), in contrast to developed economies.
   e. *Lack of exceptionally light 'edge' ICT solutions*: following on from the latter, the ICT solutions at the edges must be 'very light' in contrast to those in developed markets.

3. **Policy, regulatory and leadership failures**
   a. *Lack of awareness & policy/regulatory failures*: lack of awareness of the benefits of [and dangers of] being connected online is typically a failure of policy challenge. Policy and regulatory failures abound in low-income economies. The opportunity costs in terms of economic, developmental and social benefits are huge.

b. *Lack of creativity in realising universal connectivity*: realising universal connectivity in low-income economies requires *true creativity* in aspects like the (re)definition of 'universal' (see Chapter 3). This is a key challenge for both policy and regulation in low-income countries, which developed economies do not *acutely* face.

c. *Poor leadership, and collaboration challenges*: there is no substitute for good leadership in the resolution of 'hard' societal problems like poverty, illiteracy, etc., and lack of Internet connectivity. Low-income countries must collaborate more to seek solutions.

d. *Lack of 'home-grown' solutions or solutions that fits with the particular socio-economic needs of a country*: e.g., M-Pesa in Kenya (see Chapter 5) or Prepaid mobile in South Africa which has been a runaway success in South Africa. A 2025 TechCabal report[158] states that 80% of prepaid phones in South Africa are sold through Pepkor stores, serving more than 34 million low-income earners, which is about 63.5% of the population.

**Part II** of this book is founded on the central thesis that addressing the challenges of connecting 2.6 billion people – and meaningfully connecting at least one billion more – requires directly confronting these issues. Any lesser effort would fail to address the core tenets of the connectivity crisis.

---

[158] https://techcabal.com/2025/05/27/8-out-of-10-south-africans-buy-their-prepaid-cellphones-from-pepkor-stores/

# 4    Civilisation in a Box

'Civilisation in a Box' is an ambitious title. One of this book's peer reviewers commented "I am wary of the rather loaded term 'civilisation' – as it conjures up images of Bible-toting missionaries roaming the jungles of darkest Africa". I understand this sentiment, and I crave the reader understands my intentionality of using this term 'civilisation' is very benign. I expound on the use of this title in Section 4.2

This chapter proposes one approach [or solution] that targets the addressing of many details [not all I hasten to confess] of the three summarised *core categories* of failures that afflict low-income countries, leading to no online connectivity for a large proportion (or most) of their citizens, and/or no meaningful connectivity for them. I have enumerated them at the start of Part II of this book.

Specifically, the proposal of this chapter aims to tackle key nub issues of all three categories, namely: (i) the demand-side market failures; (ii) the supply-side market failures; and (ii) the policy, regulatory and leadership failures.

This chapter describes one proposed solution that encompasses (or is inspired by):
(i)    the 'Lever' and 'pulley' thinking.
(ii)   the 'Hub and spoke' approach to implementing the solution.
(iii)  that light ICT Solutions are needed at the Spokes of the Hub.
(iv)  a solution which redefines 'universal access' for broadband away from 'individual'.
(v)   a 'Bottom-up' community ownership trait to the solution; and
(vi)  an approach which would require much more collaboration.

A colleague of mine, Eric J. Wilson, and I titled our proposed solution 'Civilisation in a Box'[159]. This solution suggests a multi-utility approach for rural and underserved communities. I have since seen the phrase used in other contexts, e.g. in proposing ecological solutions[160].

---

[159] https://dynamicspectrumalliance.org/wp-content/uploads/2018/05/Day2-5-7_ReasonsForAChange_HNwana.pdf
[160] https://wiki.opensourceecology.org/wiki/Civilization_In_a_Box

## 4.1 Thinking and Rationale Behind Civilisation in a Box

The core thinking behind the proposal of a multi-utility approach for rural and underserved communities is the following.

Imagine a single multi-utility service approach (including electricity, Wi-Fi, mobile, water, mobile financial services) *may* be the best path to a sustainable and scalable business case for underserved communities in rural areas, particularly rural villages that I know and love, like in my native Cameroon. We imagined the following characteristics of this solution.

1. *A single infrastructure investment:* we hypothesised and envisioned one single infrastructure investment, but an infrastructure that helps provide multiple utility services to that rural village (water, electricity, connectivity via Wi-Fi, maybe even satellite connectivity to the Internet, mobile financial services, etc.). This is because the separate infrastructure investments for water, electricity, telecommunications, etc., would clearly be most unprofitable.

2. *Multiple revenue streams:* we hypothesised and envisioned multiple revenue streams for the separate utilities services in order to maximise the [chance of] profitability of the single infrastructure investment.

3. *Containerised solution that can be installed in days:* we hypothesised and envisioned a simple containerised solution that can be installed in days, rather than months.

4. *Provision of full suite of services builds Community spirit and fosters protection of Assets:* we hypothesised that providing a full suite of multiple [several] services would build Community spirit that would foster the protection of the assets – and not their theft and/or these assets being vandalised. This is truly key – the spirit of community ownership.

5. *Primary focus on renewable energy, slashing expensive kerosene and diesel costs:* our initial primary focus was renewable energy, to reduce the cost of supplying diesel and kerosene.

6. *High Profile Projects for Corporate Social Responsibility (CSR) or even for Preferential Financing:* we envisaged and hypothesised that such a project may (or would) be both high-profile and impactful enough to tempt major corporates in various countries to fund using their CSR non-profit arms. Perhaps, preferential financing could also be attracted to such projects.

As one of the peer reviewers of this book sagely observed, such a 'civilisation in a box' approach requires a 'whole of society' / 'whole of Government' approach, and the breaking down of ministerial silos. I could not agree more.

## 4.2 'Civilisation in a Box' – A Vision for Rural Inclusion

This section describes our vision of 'Civilisation in a Box' and how it addresses rural community inclusion. Figure 4.1 summarises the business case concept in a hopefully simple and graphical way.

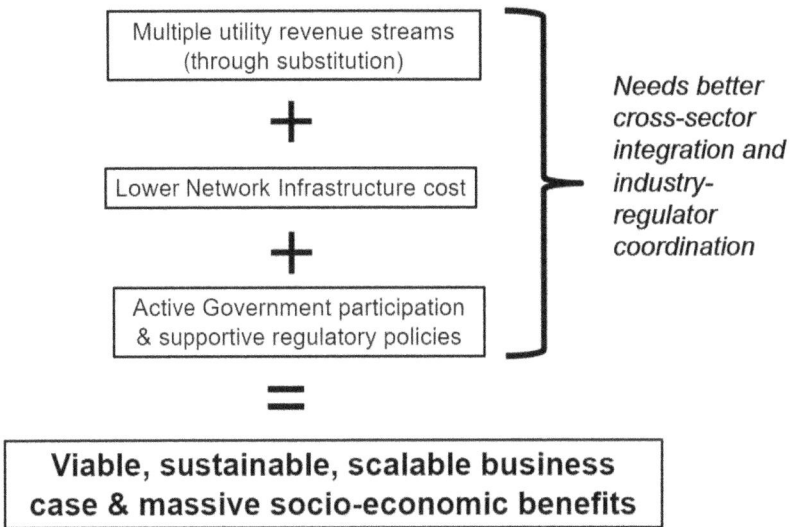

Figure 4.1 – Civilisation in a Box Business Case **Concept** for Rural and Underserved Communities (Source: Eric Wilson & H Sama Nwana)

The true vision of Civilisation in a Box is one of a holistic, multi-utility approach for underserved and/or rural communities. Specifically, imagine delivery to a village of a *single container Remote Power Unit (RPU)* installed in 2-3 days, equipped with:

o Solar panels and batteries for electric power;
o A "mini-grid" providing electricity for school, hospital, street lighting, power base stations for 2G voice (see Figure 3.3) or WhatsApp voice;
o Network connectivity for Internet access (Wi-Fi hotspots, and low cost 2G/4G);
o (Free) Terrestrial or Satellite TV for news, current affairs, education and entertainment;
o Ability to drill a well and use an electric pump to pump out clean water, enabling constant and ready water supply;
o Provision of a Hub [cf. Hubs and spoke model] with Cache servers for relevant local IT content (e.g. Mwabu education content [see Figures 3.5 and 3.6]), Government services such as IDs, social grant applications, social grants payments, etc.;
o Financial services through mobile money;
o Integrated kiosk for battery re-charging services, use by mobile money agents, provision of solar lanterns & Wi-Fi access vouchers (and cold drinks).

These are all main pillars of modern society, delivered to the village or to the rural area!

We named the proposition 'Civilisation in a Box' because it seeks to address the lack of the principal pillars of a modern interconnected society which are mostly missing in many rural areas in rural and semi-rural Africa and elsewhere in the world like most of PNG (see Chapter 2): access to reliable power/electricity; affordable access to Internet & mobile network; easily accessible clean water; basic financial inclusion and financial services, and more.

The consequences of lacking these are significant, as outlined below.

### 4.2.1 The Poorest in Society Pay the Highest Price for These Gaps – and Watch Out for GINI

Perversely, the reality is that the poorest in society pay the highest price for these basic modern-day civilisation 'gaps' in such poor rural areas:

- High direct costs for basic utility necessities (e.g., high cost of kerosene, diesel, travelling to places with access, etc.);
- Very high opportunity costs: the opportunity costs in terms of time and productivity are humongous (e.g. hours for youngsters to fetch water, travelling to major cities to have access to basic goods and services, etc.);
- Health and disease, poor education, lack of modern services and much more.
- High connectivity costs (GNI vs. GINI): entry-level mobile data plan cost for those living in rural areas could be anywhere between 11-25% of monthly Gross National Income (GNI). GNI is averaged out and does not consider GINI coefficient[161] – or Gini Index[162] – which is a numerical measure of income or wealth inequality within a population. It demonstrates how wealth is sometimes highly concentrated in a small segment of the population. For example, South Africa has the highest GINI coefficient in the world as of 2025[163] at 63 with a GNI of USD 6,530 (2023) – i.e., high income and *extreme inequality*. This is compared to Rwanda's GNI (2023) of USD 940 and a GINI of 43.7[164], depicting both low income but *moderate inequality*. A country with a Gini of 0.30 has low inequality – most people earn roughly similar amounts. A country

---

[161] GNI vs. GINI: Understanding the distinction between GNI and the Gini coefficient is essential for policymakers. Gross National Income (GNI) measures the total income generated by a country's residents and businesses, including earnings from abroad. However, as an average, GNI does *not* reflect how income is distributed within the population. The Gini coefficient addresses this limitation by quantifying income inequality on a scale from 0 (complete equality) to 1 (complete inequality) – or 0 to 100%. Consequently, two countries with identical GNI per capita may experience vastly different socioeconomic conditions due to disparities in their respective Gini coefficients: one may have a robust middle class, while the other may face pronounced gaps between affluent and impoverished groups.

[162] GINI is not actually an acronym. It is named after the Italian statistician and sociologist Corrado Gini, who introduced the concept in 1912.

[163] https://worldpopulationreview.com/country-rankings/gini-coefficient-by-country

[164] *Ibid.*

with a Gini of 0.60 has high inequality – a small elite earns much more than the rest.

## 4.2.2 These Basic Gaps Perpetuate Demographic and Political Challenges

These basic gaps perpetuate the demographic (and political) challenges facing many LDC, LMC, SIDS, LMIC and LLDC nations:

- Mass migration of the young and skilled to urban areas (creating infrastructure stress in the cities and urban poverty)[165];
- Inability to attract businesses and jobs to rural areas[166];
- Inability to attract skilled professionals (teachers, doctors, nurses, etc.)[167];
- Low school attendance (time spent fetching water, lack of incentive)[168];
- High birth and mortality rates[169];
- Disaffection and social/political unrest (Resnick, 2020).

These issues are real in many countries. Despite decades of good intentions on the part of scores of developmental organisations mentioned in Chapter 1 (Section 1.5), billions of people remain in underserved communities – and they increasingly have no choice but to migrate to already populated LMC and LMIC cities.

---

[165]https://www.un.org/development/desa/pd/sites/www.un.org.development.desa.pd/files/unpd_egm_201709_s3_paper-awunbila-final.pdf & https://www.iom.int/sites/g/files/tmzbdl486/files/our_work/ICP/MPR/WMR-2015-Background-Paper-MAwumbila.pdf

[166] World Health Organization (2022). *Health Workforce Status in the WHO African Region: Findings and Action* - https://gh.bmj.com/content/7/Suppl_1/e008317 & Makgamatha & Moikanyane (2019). *Insufficiency of Government Interventions in Promoting Rural Enterprises in SA*. University of Limpopo

[167] https://gh.bmj.com/content/7/Suppl_1/e008317

[168] https://www.mdpi.com/2673-995X/2/2/10 & https://journals.sagepub.com/doi/pdf/10.1177/1745499916632424

[169] https://www.who.int/news-room/fact-sheets/detail/newborn-mortality & https://bmcpublichealth.biomedcentral.com/articles/10.1186/1471-2458-9-462

## 4.3 Mauritania: The First Step in 'Civilisation in a Box'

The Civilisation in a Box proposition did not only end up as a paper exercise. Significant aspects of it were implemented, though admittedly not all of its constituent parts were.

### 4.3.1 The Nimjat Implementation

Figure 4.2 summarises for the reader the Civilisation in a Box Vision that underpinned the implementation in Nimjat, Mauritania.

**Vision: Civilization in a Box**

*Holistic, multi-utility approach for underserved communities*

Imagine delivery to a village of a **single container (RPU)** installed in 2-3 days, equipped with:

- solar panels and batteries for electric power
- "mini-grid" provides electricity for school, hospital, street lighting, BTS
- Network connectivity for internet access (Wi-Fi hotspots, 2G/4G)
- (Free) Terrestrial or Satellite TV for news, education and entertainment
- Ability to drill a well with electric pump for clean water supply
- Cache servers for relevant local IT content (Education, Agriculture)
- Financial services through mobile money
- Integrated kiosk for battery re-charging services, mobile money agent, solar lanterns & Wi-Fi access vouchers (and cold drinks!)

➔ **All main pillars of modern society are delivered to the village!**

Figure 4.2 – Civilisation in a Box Business Vision for Rural and Underserved Communities
(Source: Eric Wilson & H Sama Nwana)

Mauritania is a country located in North West Africa, bordered by Western Sahara to the north and northwest, the Atlantic Ocean to the West, Algeria to the northeast, Mali to the east and southeast, and Senegal to the southwest. By land area, Mauritania is the 11th largest country in Africa but *only* has a population of circa 4.3 million people [as of 2024]. This means Mauritania has a population density of just circa 5 people per km$^2$ (or 13 people per square mile) – extremely low compared to the world population density of 60.29 per km$^2$ according to Macrotrends[170]. This is one of the core reasons it is not economical to online-

---

[170] https://www.macrotrends.net/global-metrics/countries/wld/world/population-density

connect Mauritanians. The village of Nimjat is a remote one in Mauritania, located about 150 kilometres south-east of the capital, Nouakchott, with a population of only around 1,500 people [as at 2019].

## Civilization in a Box

Winch Energy Offgrid solution in Mauritania

THE RPU PROVIDES BETTER SANITATION AND **ACCESS TO WATER** BY POWERING ELECTRIC PUMPS

Figure 4.3 – Civilisation in a Box Business Implementation in Nimjat, Mauritania
(Source: Winch Energy[171], reproduced with permission of the Owner, Nicholas Wrigley)

Winch Energy[172] was the key stakeholder in making a reality of the vision of Figure 4.2 in Nimjat. Winch Energy was/is a renewable energy developer who implemented a solar-based electricity mini-grid system in Nimjat, providing electricity to the local school, dispensary, mosque, 70 houses, as well as streetlights – as shown in Figure 4.3. This has been transformative to the village which had never had any grid-based electricity before, albeit a mini-grid. The mini-grid is powered by an innovative, *containerised PV* (photovoltaic) *solar system* with battery backup – constituting a Remote Power Unit (RPU). A containerised PV solar system is a portable and modular solar power solution housed within a shipping container as shown in Figure 4.3 (top left). PV systems are designed for quick deployment and flexibility, making them ideal for remote

---

[171] https://www.winchenergypower.com/about-us/
[172] *Ibid.*

locations, emergency response, military use, and for other applications where a fixed installation is infeasible. The modular system came with three different [power] models available at the time: 7kW, 17kW, 30kW, which could be scaled to quickly meet growing demand. The system was assembled in 2-3 days and could be installed using local labour. Winch Energy started with an RPU 17kW installation but quickly had to scale it up because of demand.

Figure 4.4 – The Winch Hub
(Source: Winch Energy[173], reproduced with permission of the Owner, Nicholas Wrigley)

Overall, the PV solar-based system delivers reliable AC supply via the mini-grid in the village, with batteries inside the container guaranteeing a reliable power supply on a 24/7 basis – acting as a back-up for night-time and during adverse weather conditions. Each RPU came with the option of a 55-inch satellite TV installed on the side of the container, and satellite broadband along with a Wi-Fi router that provides broadband at 3G mobile speeds.

Since batteries were/are fully charged by circa midday in Mauritania due to the extremely high irradiation levels in the Sahara Desert area, it was decided to use the 'excess' energy to pump water from a drilled well, thereby providing clean water supply.

The Winch Hub shown in Figure 4.4 is the Internet Hub and Battery Charging unit arranged in a bespoke containerised kiosk. This allows for low-income

---

[173] https://www.winchenergypower.com/about-us/

households to connect to the Internet, and from where they can lease batteries to run household appliances. The leasable battery is designed with USB ports to power small appliances such as LED lights, fans and mobile charging.

### 4.3.2 Impact to the Village

Hopefully, as Figure 4.3 further illustrates, the project has significantly improved the quality of life in Nimjat by providing – not only – reliable electricity power, but Wi-Fi access, DStv[174], electric-pumped clean water, a first-ever dispensary in the village housing live vaccines, etc.

The village leadership informed us/Winch Energy that this was the first time they had ever had electricity and other utilities in the village, and noted that they are expecting some significant migration back to the village as a result. There is outstanding community support for the project which helps foster a protective attitude for these RPU-based village assets.

## *4.4 How Civilisation in a Box Meets the Characteristics Set Out in Chapter 3*

I hope it is already truly clear to the reader how this Civilisation in a Box proposal implements the six essential traits of Chapter 3, i.e.

(i)    the 'Lever' and 'pulley' thinking;
(ii)   the 'Hub and spoke' approach to implementing the solution;
(iii)  light ICT Solutions at the Spokes of the Hub;
(iv)  which redefines 'universal service' for broadband away from 'individual';
(v)   'Bottom-up' community ownership of the solution; and
(vi)  provides opportunities for collaboration with others to 'productise' the civilisation-in-a-box concept.

---

[174] DStv is a digital satellite television service that offers a wide range of entertainment options, including live sports, TV shows, movies, and kids' programming. It is available in several African countries and provides various packages to suit different preferences and budgets. Source: www.dstv.com

Table 3.2 illustrates <u>how</u> the Civilisation in a Box Solution in Nimjat, Mauritania abides by the characteristics elaborated upon in Chapter 3.

| | Essential Characteristic | Examples (non-exhaustive) of how Civilisation in a Box Solution in Nimjat, Mauritania – abides by the characteristic |
|---|---|---|
| 1 | The 'Lever' and 'Pulley' thinking - bringing *'Civilisation' to the Village,* using a single RPU-based Infrastructure | (i) Lever/Pulley 1 – the RPU-based Infrastructure's mini-grid is/was *providing electricity to at least 70* homes. It uses renewable Solar Power – instead of unaffordable Grid Electricity or Diesel (cf., see Table 3.2) |
| | | (ii) Lever/Pulley 2 – the <u>same</u> RPU-based Infrastructure's mini-grid is/was providing clean water pumped from a drilled well, which can be fetched reliably and from the 'Hub' centre of the village – a key essential utility of life. |
| | | (iii) Lever/Pulley 3 – the <u>same</u> RPU-based Infrastructure's mini-grid is/was providing broadcasting and telecommunications services like DStv and Wi-Fi respectively – thereby connecting many before-unconnected villagers. |
| | | (iv) Etc. |
| 2 | The 'Hub and Spoke' Approach | (i) The containerised PV solar system-based RPU is clearly a 'Hub' that powers the entire mini-grid connected to 70 'Spoke' homes – as depicted Figure 4.3. |
| | | (ii) The Winch Hub shown in Figure 4.4 is a clear Internet Hub, also providing for battery charging – and which allows for low-income household to connect to the Internet - and from where they can lease batteries to run household appliances. |
| | | (iii) To manage the demand and production loads of the mini-grid, an automated and remote |

| | | | |
|---|---|---|---|
| | | | power management system at the Hub RPU is/was implemented using smart meters installed at the 'spoke' households. |
| 3 | Light ICT Solutions at the Spokes | (i) | Due to the inherent limitations of the power/electricity capacity at the Hub (7kW, 17kW, 30kW, etc.) and the use of smart meters, 'power hungry' demand at 'spoke' homes are discouraged and/or rationed. |
| | | (ii) | The Winch Hub model shown in Figure 4.4 offers opportunity for local villagers to lease batteries, mobile phones for making calls and other small appliances – all Light ICT solutions. |
| 4 | Redefinition of 'Universal Service' | (i) | Though not specially intended at the time, the RPU-based services implicitly define and implement 'Community'-based (cf. Section 3.4) *universal services* through providing various services from the Hub, e.g. DStv, Wi-Fi or even access to clean water. |
| | | (ii) | Following on from the latter, the Hub RPU can also enable other 'universal services' defined as 'for every home', and not for 'every individual', i.e., services at the Hub, e.g. Wi-Fi-based broadband. |
| 5 | Empowered, 'bottom-up', 'Community-led' Ownership | (i) | This project was community-led. The Nimjat Civilisation in a Box solution was led, commissioned and largely paid for by the enlightened Local Chief of the 1500-people village who had personally 'felt the pain' of losing so many services from his village with substantial migration of youth away from his village. He hypothesised that by bringing civilisation to his village, he could stem this migration tide and even reverse it. |
| | | (ii) | The Chief was right. Since the project was concluded, young tailors, seamstresses, barbers, shop owners and even retirees who |

| | | |
|---|---|---|
| | | had migrated away from the village have been returning. |
| 6 | Collaboration with others in order to 'productise' the civilisation-in-a-box concept | (i) This project was *not* only community-led, but it involved much cooperation. Led, by the Local Chief, it involved much local cooperation amongst groups and professionals in the village: electricity technicians, women's groups, mosque leadership, local Government authorities, health professionals, etc. – and of course Winch Energy from the UK who supplied the multi-utility infrastructure. |
| | | (ii) The water well happened due to the involvement of UK-based Charity, Water Aid[175]. |

Table 3.2 – Illustration of <u>how</u> Civilisation in a Box Solution in Nimjat, Mauritania
– abides by the characteristic of Chapter 3

## 4.5 The Nimjat Experiment Remains Incomplete

Lest I leave the reader with the impression that the Civilisation in a Box concept [and implementation] in Nimjat, Mauritania is a runaway success, I should clearly point out here that the jury is still out. The 'experiment' remains incomplete.

The primary reasons are outlined below:

a. *Expensive Installation*: the entire Nimjat (Mauritania) implementation of the Civilisation in a Box concept is still to be <u>properly</u> productised, cost-optimised and replicated elsewhere. It was a relatively expensive *one-off* installation.

b. *Totally Unproven Business Model*: Figure 4.1 depicts what my colleague Eric Wilson and I posit to be the core business case of the 'Civilisation in a

---

[175] www.wateraid.org – they help people break free from poverty and change their lives for good through the three essentials of clean water, decent toilets and good hygiene.

Box Business' case concept for Rural and Underserved Communities. We posit/posited that

Multiple utility revenue streams (through substitution) **✚**
Lower Network Infrastructure costs (through sharing) **✚**
Active Govt Participation & Regulator Support

**=**

*Viable, sustainable, scalable business case with massive socio-economic benefits*

This business model above remains completely unproven. I would be so honoured to see these couple of reasons fully addressed by several other experiments elsewhere. This is the core reason why I suggest in Chapter 3 that that any successful approach requires another especially important characteristic or attribute: deep collaboration and cooperation amongst like-minded LMIC/LMC countries and unconnected communities.

c. *A Limited Supply-side Intervention*: ultimately, Nimjat is/was a supply-side intervention, albeit in the context of pent-up demand. It did not extend to other demand-side interventions, e.g. training teachers to use computers and paying them to do so.

As the proverb goes – 'A problem shared is a problem halved'. I recently came across another variation to this proverb, which is arguably more apt to what I am articulating here: *"a problem shared with the right person (or people), at the right time and with the right intentions, is a problem halved"*[176]. I hope that leading LMIC, LMC, LIC, LDC countries, and EDC countries like India will collaborate with each other and major development agencies (see Section 7.4 and Section 1.5) to bring billions of their citizens online sustainably. They seem to be the right people at the right time (2025) with the right intentions.

---

[176] https://www.linkedin.com/pulse/problem-shared-halved-think-again-olajumoke-ola

## 4.6 Evaluating Civilisation in a Box vs. Connectivity Barriers

How do I assess 'Civilisation in a Box' against the summarised core categories of reasons why the tail 2.6 billion remain unconnected? At the start of Part II of this book, I provide the three *summarised core categories* of reasons <u>why</u> the 'Tail 2.6 billion' are *not connected* to the Internet as of 2025 (cf. Figure 1.1) – as well as <u>why</u> a significant fraction of the 'Mid-Tail 2.7 billion' are *not meaningfully connected*, namely:

1.  *Demand-side market failures*: crushing poverty; poverty (low income); lack of awareness, connectivity being a low-priority concern; illiteracy; lack of digital skills; low population density (rural areas); usage-based pricing models, poverty, miscellaneous issues, etc.

2.  *Supply-side market failures*: no electricity/power; poor basic infrastructure like roads and rail; unviable [infrastructure] business models; lack of suitable 'hubs and spokes' ICT solutions; lack of exceptionally light 'edge' ICT solutions, etc.

3.  *Policy, regulatory and leadership failures*: lack of awareness & policy/regulatory failures; lack of sufficient creativity in realising universal connectivity; poor leadership and lack of collaboration challenges; the need for 'home-grown' solutions like M-Pesa in Kenya.

So, how does the civilisation in a box proposition fare in meeting these summarised categories of challenges? My *brief* assessment is neither scientific nor objective, but here it follows.

First on supply-side market failures, it appears that a civilisation in a box-type proposition and approach would largely address the type of supply-side market failures that typically afflict LICs and LMICs. Second on policy, regulatory and leadership failures, I hope the reader would concede that the proposition – accompanied by good leadership and creativity in how universal connectivity is defined in policy/regulation – would go a long way long to addressing the long tail 2.6 billion challenge. The proposition certainly is also 'home-grown' – borne of the realities of low-income countries/economies. Lastly, I concede the proposition struggles with the demand-side market failures. It will never address hard problems like poverty, affordability, illiteracy and lack of digital skills. These are societal challenges that Governments must lead on. However, the proposition *implicitly* 'acknowledges' demand-side challenges like low population density of rural areas, as well as would address usage-based pricing models with technical solutions like Wi-Fi.

## 4.7  Summary

This chapter has introduced the 'civilisation in a box' concept, vision and business model – and an incomplete implementation in Nimjat, Mauritania. I still posit that the concept, vision and business model could be *revolutionary* towards enabling the connecting a good proportion of the long tail 2.6 billion unconnected – and meaningfully connecting hundreds of millions more.

# 5   Democratising Connectivity for the Unconnected 2.6 Billion

Democratising and localising connectivity responsibility of the 2.6 billion not online is an absolute must – I posit. I argue in this chapter that connecting most of the tail 2.6 billion unconnected will require, from LMICs and LICs, true 'home grown' solutions – like the proposal of the last chapter portends to. To allow such solutions to emerge, I further posit that these Governments would need to 'democratise' [as much as possible] and localise much policy, regulatory, financing, technical challenges and more. Most importantly, also democratise *responsibility* – to *local* and *non-traditional* 'telecoms' stakeholders in society including community NGOs, local schools, local universities, local churches, local supermarkets, local businessmen and businesswomen, etc. Essentially, we would also need to evolve and licence a new 'breed' of Internet Service Providers (ISPs) to help connect the poor in these LIC, LMIC and LDC economies. The current traditional MNOs and fixed operators have patently failed to do so – and will continue to fail.

It is true that ISPs must be profitable; however, this chapter rails against regulatory frameworks that effectively prevent small ISPs from entering the market. This phenomenon is evident in numerous developing countries.

Some may argue that such a proposal on democratisation and localisation of Internet *connectivity responsibility* is hardly 'out of the box' thinking. Fervently, I beg to disagree. The prevailing culture in most LDCs, LMICs and LICs is one of 'top-down' (from the centre) and Government-knows-best, i.e. Big Government. Despite challenges in meeting their Sustainable Development Goals (SDGs) and universal digital connectivity recommendations, governments often hesitate to grant local villages and towns the authority to address these issues independently. These communities, which face significant difficulties, could greatly benefit from having more control over these efforts. I observe this culture in LMIC/LIC after LMIC/LIC across the globe, wherein bureaucrats[177] in their capital cities intervene and stop local initiatives – particularly in Sub-

---

[177] Ministers, Regulators, Policy makers, State-appointed Governors, etc.

Saharan Africa (SSA), and even most particularly in the seventeen (17) former France SSA colonised countries[178]. I have observed this Big Government culture closely in my own country of origin, Cameroon, and it prevails in practically all of the fifteen-to-twenty (or so) LDC/LIC/LMIC countries I have been visiting annually for more than a decade. I argue in this chapter that such Big Governments must largely get out of the way of more local solutions to their portion of the tail 2.6 billion unconnected challenge.

In this chapter, I explain why democratised 'home grown' solutions are key, and why changing the Big Government connectivity culture with LDCs, LICs and LMICs is even more critical. This would be quite revolutionary and 'out of the box' for many of these Governments.

**Caveat**: I acknowledge that such bottom-up and decentralised solutions as I am advocating in this chapter require guard-rails. The intent of this chapter is to foster an environment that encourages diverse and innovative approaches to connectivity, while avoiding excessive regulation that could hinder promising connectivity innovations. Let a thousand 'connectivity' flowers bloom!

## 5.1   M-Pesa: the Kenyan 'Home-Grown' Solution

At the end of the last Chapter [Chapter 4], I make mention of why the civilisation in a box proposal is *'home grown'*[179] – borne of the realities of digitally-connecting the poor who inhabit low-income economies. Earlier in this book, I refer to Kenya's M-Pesa, even though I do not elaborate on what M-Pesa is.

[Safaricom's[180]] M-Pesa is a mobile phone-based digital wallet and money transfer service widely used in Kenya and elsewhere in Africa like Democratic Republic of Congo (DRC), Egypt, Ghana, Kenya, Lesotho, Mozambique,

---

[178] Benin, Burkina Faso, Cameroon, Central African Republic, Chad, Comoros, Republic of the Congo, Côte d'Ivoire, Djibouti, Gabon, Guinea, Madagascar, Mali, Mauritania, Niger, Senegal and Togo

[179] It is sometimes lost that M-Pesa is not entirely 'home-grown'. It was based on an idea from a Vodafone (UK) employee. M-Pesa's unique fit to the local needs of Kenya made the idea a runaway success.

[180] Safaricom is not only the biggest mobile network operator in Kenya, but in all of East Africa – https://www.safaricom.co.ke/about/who-we-are/our-story

Ethiopia and Tanzania[181]. Nowhere in the TMT sector worldwide is the aphorism "necessity is the mother of invention" truer than with mobile money.

In 2020, it was reported by the Central Bank of Kenya (CBK) data that M-Pesa mobile money transactions accounted for more than 50% of Kenya's GDP[182]. This means Kenyans transacted half the equivalent of Kenya's GDP through their mobile phones' digital wallets in 2019/20. According to more recent data from the Central Bank of Kenya[183], as of 2024, mobile money transactions in Kenya – including M-Pesa – accounted for 53% of Kenya's GDP. This reflects the continued dominance of mobile money in the country's economy, highlighting its pivotal role in financial inclusion and economic activity.

According to Safaricom's May 2023 Investor Presentation, the company:

"grew our 30-day active M-PESA users by 7.2Mn in the last three years. Grew M-PESA chargeable transactions / customer from 12.9 to 23.5 the last three years" (Source: [184]).

And this is why Kenya's Safaricom is now the biggest banking institution in the country as of 2024 too. Indeed, as of September 2023, M-Pesa had a 97% share of m-money subscriptions in Kenya, while Airtel Money has 2.9%, and T-Kash has just 0.1%[185].

M-Pesa's dominance continues to grow.

"Consumer payments were the largest revenue contributor, accounting for 62.9% of M-Pesa's total revenue, amounting to KES 48.6 billion. This reliance on consumer payments illustrates how embedded M-Pesa is in people's routines, from buying essentials to paying bills. The consistent rise

---

[181] M-PESA Africa - https://www.m-pesa.africa/what-is-mpesa
[182] https://www.paymentscardsandmobile.com/mobile-money-transactions-half-of-kenyas-gdp/
[183] https://bitcoinke.io/2025/03/mobile-money-agents-in-kenya-in-2024/
[184] https://www.safaricom.co.ke/images/calendars/FY23-Investor-Presentation-11-May-2023.pdf
[185] https://www.telcotitans.com/vodafonewatch/rumours-linking-safaricoms-m-pesa-outage-to-kenyan-tax-agency-denied/7651.article

in consumer transactions reflects how M-Pesa continues to be a trusted tool for managing personal finances" [186].

M-Pesa's turnover has become a significant part of the Kenyan economy. In 2023, M-Pesa transactions made up nearly 59% of Kenya's GDP[187]. This indicates that a large portion of the country's economic activity passes through M-Pesa, illustrating its role in Kenya's financial system. In the 2024 financial year, M-Pesa contributed to over 42% of Safaricom's annual revenue, an increase of nearly 10 percentage points from 2021[188]. This reflects Kenya's position as a major mobile money market and highlights the growing importance of mobile money for Mobile Network Operators (MNOs).

As noted, 53% of [Kenyan] GDP was transacted by mobile money agents in Kenya in 2024[189]. Who would have thought that twenty years ago?

*Why am I recounting the M-Pesa story in the context of bridging the digital divide?* The Kenya Safaricom/M-Pesa story is a textbook example of a 'home-grown' solution – a telecoms company officially launched and licensed to provide mobile telecoms services in October 2000 has now evolved to become the most important financial institution in the country, transacting 59% of the country's GDP in 2023. M-Pesa would never have happened in developed economies. They do not *need* it, with practically all citizens in developed countries having a bank account in addition to credit/debit cards.

Indeed, as I was concluding the peer review corrections of this book, Vodafone Group announced that Safaricom's mobile money platform M-Pesa generated KES 161 billion in Kenya's service revenue for the year ending March 31, 2025[190], accounting for a whopping 44.2% of Safaricom's total service revenue in Kenya[191].

[186] https://www.abojani.com/safaricoms-h1-2025-results-mpesa-impact/
[187] https://www.wbs.ac.uk/news/how-mpesa-cornered-the-market/
[188] https://www.gsma.com/sotir/wp-content/uploads/2025/04/The-State-of-the-Industry-Report-2025_English.pdf
[189] https://bitcoinke.io/2025/03/mobile-money-agents-in-kenya-in-2024/
[190] https://vodacom.com/news-article.php?articleID=15684
[191] https://www.the-star.co.ke/news/2025-05-09-m-pesa-drives-safaricoms-record-sh388-billion-revenue

Mobile money is now a phenomenal revolution beyond Kenya and across Sub-Saharan Africa. Indeed, according to GSMA's 2025 Mobile Money Report[192],

> "In Benin, Côte d'Ivoire, Ghana, Guinea, Guinea Bissau, Senegal and Liberia, mobile money contributed more than 5% to GDP. In East Africa, mobile money contributed more than 5% to the GDPs of Kenya, Rwanda, Uganda and Tanzania. Elsewhere in Sub-Saharan Africa, mobile money's contribution to GDP has been mixed. In Central Africa, Cameroon, Congo and Gabon each saw a contribution between 5% and 8%", p.25.

A key reputable 2017 study found that access to M-Pesa lifted approximately 2% of Kenyan households (around 194,000 families) out of extreme poverty[193] – this figure would have increased by now in 2025. The study found that this was particularly impactful with female-headed households, which saw greater increases in consumption and economic activity/stability.

## 5.2 Transforming Connectivity Culture in LICs: Insights from Brazil

I posit the transformation and/or changing the connectivity 'cultures' in LIC economies through learnings from UMIC Brazil. We really and drastically need to change the prevailing culture in LDCs, LICs and LMICs that connectivity to the Internet must be provided by one or more of 'big boy' MNOs. It is both instructive and illustrative to learn some lessons from UMIC Brazil.

The role of Wireless Internet Service Providers (WISPs) in Brazil in closing the digital divide cannot be overstated. There is much for other LICs, LMICs and LDCs to learn from UMIC Brazil as regards the supply of Mobile Broadband Services.

Some key lessons from Brazil include the following.

---

[192] https://www.gsma.com/sotir/wp-content/uploads/2025/04/The-State-of-the-Industry-Report-2025_English.pdf
[193] https://www.cgap.org/blog/why-does-m-pesa-lift-kenyans-out-of-poverty

a.  *Improved Internet connectivity*: in recent years, Brazil has experienced improved Internet connectivity thanks to new entrant mobile broadband players, Wireless Internet Service Providers (WISPs[194]).

b.  These WISPs have employed *unlicensed radio spectrum* that allows for easier deployment of Wi-Fi networks, which has significantly contributed to bridging the digital divide in Brazil. The Brazilian Government has been leading the efforts in providing Internet access in its underserved areas. In Brazil, such new entrant WISPs have played a crucial role in providing mobile broadband service to low-income citizens.

c.  "WISPs tend to serve predominantly *lower income groups households*: according to the CeTIC.br survey, most of the WISP customers *belong to the C, D, and E strata. In these segments, wireless technology is hardly substituted by other kind of access technology.* As a result of the lower income population concentration, *WISP Wi-Fi lines are frequently shared among neighbours.*[195] (*my emphases*).

d.  As depicted [in Table 5.1], in 2023, 16.37% of households in Brazil overall accessed broadband *by sharing a fixed broadband connection* with a neighbour. Indeed, it is much higher at circa 25% for Strata D&E. In 2023 as shown, the number of households (HHs) served by Brazilian WISPs was some 1,678,055 HHs, with a high concentration of the latter number in rural areas[196].

---

[194] A WISP (Wireless Internet Service Provider) is an Internet service provider that uses wireless networking technology to provide Internet access. Instead of relying on traditional wired connections like fixed lines, cable or fiber, WISPs use wireless communication methods, such as Wi-Fi, to connect users to the internet and/or for backhaul.

[195] https://dynamicspectrumalliance.org/2024/Assessingtheeconomicvalue6GHzBandBrazil2021-2034.pdf

[196] *Ibid.*

| Social Segment | Sharing Connections | No Sharing | Fixed Broadband Adoption | Households | Sharing (% households) | Sharing (% FBB connections) |
|---|---|---|---|---|---|---|
| A | 8,765 | 1,052,661 | 976.063 | 1,083,153 | 0.83% | 0.90% |
| B | 1,299,672 | 12,246.944 | 11,579,739 | 14,114,562 | 9.47% | 11.22% |
| C | 5,002,919 | 27,353,079 | 24,111,755 | 35,864,644 | 15.46% | 20.75% |
| D+E | 4,089,649 | 12,302,864 | 9,670,733 | 24,849,111 | 24.95% | 42.29% |
| Total | 10,401,005 | 53,135,548 | 46,338,290 | 75,911,470 | 16.37% | 22.45% |

Table 5.1 - Brazil: Connection sharing of fixed broadband (2023)
Source: Dynamic Spectrum Alliance[197]

e.  According to this reputable blog[198], the

>   "Brazil Internet access industry is quite unique: *Brazil has over 20,000 ISPs* but only 3 have country-wide coverage and *40% of the companies have up to 5 thousand customers*. It is a highly competitive and heterogenous marketplace, driven by small and medium regional companies. As of December 2022, *regional ISPs account for over 50% market share in Brazil.*" – (*my emphases*).

f.  *Cost Efficiency*: as illustrated by the Brazilian example, using unlicensed spectrum reduces costs for WISPs, as they do need to bid for or pay for high IMT spectrum licenses. Some of the bigger ISPs evolve to start bidding for some regional IMT spectrum licenses.

g.  *Economic Growth*[199]: enhanced connectivity – particularly to rural and unconnected areas – obviously drives economic growth by enabling access to digital services, education, healthcare and more.

---

[197] https://dynamicspectrumalliance.org/2024/Assessingtheeconomicvalue6GHzBandBrazil2021-2034.pdf
[198] https://wifinowglobal.com/news-and-blog/focus-on-brazil-biggest-challenge-for-isps-is-to-deliver-great-wi-fi-inside-all-rooms-of-the-house-says-abrint/
[199] https://www.kictanet.or.ke/challenges-of-spectrum-access-in-africa/

h.  *Wi-Fi Offload for Mobile Network Operators*[200]: the practice of Wi-Fi offloading allows mobile networks operators to offload traffic from their congested networks onto Wi-Fi networks when possible, freeing up their valuable spectrum. This would usually improve overall network performance and reduce congestion and improve quality of service.

## 5.3   Democratising and Localizing Connectivity for the Next 2.6 Billion

What the examples of the two previous sections (M-Pesa and the Brazilian ISP market] clearly indicate is that the most optimal solutions exhibit the following 'democratic and local' traits:

(i)  *Address market failures with Brazilian low-income communities*: in 2023, the number of households (HHs) served by Brazilian WISPs was some 1,678,055 HHs, with a high concentration of the latter number in rural areas[201]. There, they cater to the Internet connectivity needs of 'lower income groups households' communities, i.e., the *C, D, and E strata* in Brazil, as shown in Table 5.1. This is clearly consistent to the entire ethos of this book. In a similar vein, M-Pesa is known to have had a significant impact on poverty reduction in Kenya (see Section 5.1).

(ii) *Home grown* – the M-Pesa story of Section 5.1 could only happen in a then LIC (and now LMIC) economy like Kenya. The Brazilian heterogeneous ISP market is uniquely home grown too.

(iii) *Local* – M-Pesa is local 'as in local to Kenya' (and increasingly across East Africa and beyond). The Brazilian WISPs have evolved a 'local' [as in 'community'] characteristic, where they address the needs of 'lower income groups households' communities.

(iv) *Democratic*: this means 'bottom-up'-led solutions to societal problems are also enabled in society, rather than preserving and protecting just top-down solutions. As we see above, the Brazilian Internet provision sector uniquely

---

[200] https://www.antlabs.com/blog-articles/why-wifi-offloading-is-even-more-compelling-in-a-5g-driven-world/
[201] *Ibid.*

has more than 20,000 ISPs with only 3 having country-wide coverage. 40% of the companies have up to 5 thousand customers *only* – driven by SME businesses. Incredibly, their regional ISPs accounted for over 50% market share in Brazil as of December 2022. This is a highly *democratic* [20,000 ISPs with less than 5,000 customers each], competitive and heterogenous marketplace. The 3 'big boy' MNOs have less than 50% of the Internet connectivity market in Brazil. Enabling this extremely high number of ISPs has been a real coup for Brazil.

(v) *Use widespread and shared technology solutions*: the WISPs have employed *unlicensed* [open] radio spectrum that allows for easier deployment of non-proprietary and unlicensed Wi-Fi (i.e. widespread) networks. Revisit Table 5.1 which shows that 16.37% of Brazilian households in 2023 accessed broadband *by sharing a fixed broadband connection* with a neighbour. This is both clever and economically efficient, 'home grown' and democratic too.

(vi) *'Democratic' policy & regulation*: clearly, even Internet policy and regulation has been democratised in Brazil, as is evidenced by the 20,000+ ISPs in Brazil who control more than 50% of the Internet market in the country. Most other LIC/LMIC/LDC economies continue to promote an Internet policy and regulation 'culture' which largely expects just the 'big boy' MNOs to Internet-connect the citizens of their countries. Brazil shows that this is truly short-sighted.

(vii) *Innovative Solutions for Democratizing Wi-Fi in Developing Economies: Kenya's Initiative to Transform Power Transformers into Internet Hotspots*: Innovative ICT ideas emerging from Kenya should be taken seriously, particularly following the success of M-Pesa. So, I was both intrigued and frankly impressed when I heard of an initiative in Kenya to turn all its power transformers into Internet hotspots. With over 70,000 transformers across Kenya, the Government, in collaboration with the Kenya Power and Lighting Company, aims and plans to leverage this transformer infrastructure to expand Internet access. The initial goal is to establish

25,000 Wi-Fi hotspots, eventually scaling up to all 74,000 transformers[202]. The innovation proposes to use power lines rather than more expensive dug-in fibre on the ground because installing the fibre alongside Kenya Power cables would drastically decrease costs[203]. The technical details of how this approach to 'democratise' Wi-Fi in Kenya using their 74,000 power transformers is yet to be worked out. For example, I am unsure if this approach proposes to use Power-Line Communications[204] (PLC) or whether it is no more than aerial fibre strung from electricity pylons. Either way, imagine if it stimulates thousands of ISPs in Kenya like in Brazil. Other countries may want to consider such innovative ideas too. I suggest the reader now also reads this section in conjunction with Section 9.8 (Chapter 9) in which I introduce the very innovative *Connect a Billion* initiative.

## 5.4 Reducing Government Barriers in LMICs, LICs, and LDCs

LMIC, LIC and LDC 'Big' Governments need to get out of their own way of bridging the digital divide. The entire rationale for the HIC, UMIC, LMIC and LIC classification designations by the World Bank is to help 'standardise' and provide a framework and guidance for understanding the economic development levels/needs of different countries, guiding necessary policies, investments, and international aid. It would truly help if the World Bank and other International Aid organisations also promote the idea of Big Governments getting out of their own ways of locals in taking responsibility for their own perennial challenges.

At the start of this chapter, I note that some may argue that my proposal on democratisation and localisation of connectivity *responsibility* would be quite revolutionary and 'out of the box' for many of these LMIC/LIC 'Big'

---

[202] https://broadcastmediaafrica.com/2024/07/21/kenya-to-use-power-transformers-into-internet-hotspots/ or https://www.powersystems.technology/news/us-news/kenya-to-transform-power-transformers-into-internet-hotspots.html

[203] Kenya notes that Rolling out fiber by digging trenches costs us approximately KSh2.3 million (US$17,510) per km. Using the approach of rolling out fiber alongside power cables would drop the costs to KSh600 000 ($4 568) per kilometer.

[204] PLC uses existing electrical wiring to transmit data by superimposing a high-frequency signal over the standard AC power signal. PLC does work in countries that use 110V vs. 230V, as well as in 220–240V countries, like most of Europe, Asia, and Africa. In fact, many PLC technologies were specifically designed with these higher-voltage, 50Hz systems in mind

Governments. Believe me – I have tried. It is just *not* in the DNA of many of these countries to devolve some of such responsibilities (like local Internet connectivity) to local ownership. Using the UMIC Brazilian Internet market example of Section 5.2, LDC/LIC/LMIC Governments prefer the scenario of being able to 'control' their equivalent 3 'big boy' MNOs in Brazil with national connectivity coverage – than a counter scenario that cedes 'control' to a cacophony of 20,000 other much smaller ISPs with less than 5,000 subscribers each. This explains why there are only circa a total 7,000 ISPs in Low- and Middle-Income Countries (LICs/LMICs) *combined*[205] – and yet Brazil *singularly* can boast of 20,000.

I find that this prevailing 'Big Government' culture in many of these LICs/LMICs/LDCs is extremely disempowering, mitigating against locally-led solutions to so many societal challenges.

The *demand-side* challenges that I outline at the start of Part II of this book are usually the most difficult, and are typically quite acute in rural areas of LDCs, LICs and LMICs: poverty (low income); lack of awareness; connectivity being a low-priority concern to other basic necessities like clean water; illiteracy; lack of digital skills/confidence; low population density (rural areas); usage-based pricing models; poverty, etc. Note how these challenges correlate almost perfectly with/to the first seven SDG goals that I enumerate in Chapter 1:

- Goal 1: No Poverty
- Goal 2: Zero Hunger
- Goal 3: Good Health and Well-being
- Goal 4: Quality Education
- Goal 5: Gender Equality
- Goal 6: Clean Water and Sanitation
- Goal 7: Affordable and Clean Energy

I posit that these challenges require a true *co-ownership* of responsibility and/or partnership, at the very least, with locals who live in these regions. I have observed in my own village in Cameroon, as well as in other villages in Cameroon and elsewhere in Africa, where the sons and daughters of a village –

---

[205] Source: Paul Garnett, https://www.vernonburggroup.com/connect-one-billion

led by their Chief – have tried to take responsibility for some of such perennial issues like lack of sustainable clean water (SDG Goal 6) or food production cooperatives (SDG Goal 2) – or poor road infrastructure. The Government for 60 years since independence [from France] has failed to deliver on such basic commodities of life to millions of its citizens. Yet, the Government is quick to step in and 'nationalise' any such water projects, food/coffee cooperatives and other initiatives, disempowering the locals of all responsibility for their *own* challenges. Rather, they should be empowering them and giving them more resources, skills and building on their capacities. The Big Governments do nothing after disempowering the locals, and these regions continue to lack the essentials of life, leading to more migration of young people to the major cities.

I have personally tried to initiate connectivity projects in villages like mine and others, but I have been quietly [and firmly] reminded by locals that they need clean water first, and stable/affordable electrical power (SDG Goal 7). Of course, they are right. The reader can see why 'Civilisation in a Box' (cf. Chapter 4) makes sense in such contexts.

## 5.5  Summary

I cannot overstate how fervently and passionately I feel about Big Governments 'getting out of the way' more in LDC, LIC and LMIC economies and empowering locals to take (co)responsibility of perennial issues in their communities, including lack of digital connectivity. It is truly an 'out of the box' idea in many of the capital cities of these countries, though it should not be.

Peer reviewer Dr Charley Lewis – as ever – provided another insightful comment pertaining to the example successful innovations that I mention in this chapter. I think the insight is worth including here for others to reflect on:

> "The real question for me is – what unique combination of circumstances in each case allowed that model to take off and flourish? Why specifically M-Pesa in Kenya, WISPS in Brazil, prepaid in South Africa? If we can answer that question, we can pretty much solve UAS anywhere?[206]"

I leave this insight to the reader and others more expert than I am to address.

---

[206] Dr Charley Lewis, Personal communications as part of the peer reviewing of this book.

# 6 Declining Market-Driven Solutions for the Unconnected 2.6 Billion

There are increasingly no more market-led telecoms solutions to connect the tail 2.6 billion. So, in this chapter, I am audacious enough to issue my 'out of the box' heads-up *warnings* to the audiences I am targeting with this book at (cf. Section 1.5), not least the Presidents, ICT Ministers, key ICT policy makers and regulators of LDCs, LMICs and LICs.

## 6.1 Caution 1: Current Telecoms Market Won't Solve the 2.6 Billion Connectivity Gap

My first key caution (or warning) of this chapter is this: beware of waiting for solutions to the tail 2.6 billion from the current telecoms market. I articulate in summary form the 'warning' in the title of this chapter above: "Declining Market-Driven Solutions for the Unconnected 2.6 billion". Another version of the same warning could read - "the new telecoms industry innovations all ignore the challenges of connecting the tail 2.6 billion". Permit me to elaborate.

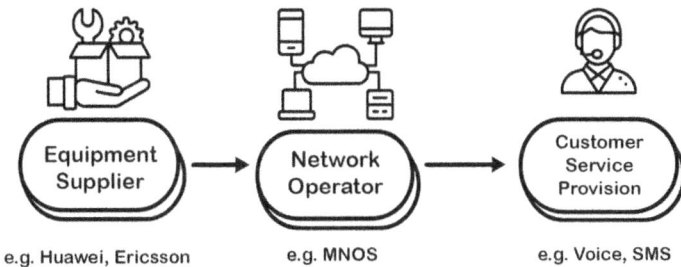

Equipment Supplier → Network Operator → Customer Service Provision

e.g. Huawei, Ericsson          e.g. MNOS          e.g. Voice, SMS

Figure 6.1 - Pre-4G/Data Age (i.e., 2G/3G) Typical Mobile Value Chain

The telecoms, media and technology (TMT) industry that I describe in much detail in Nwana (2014) has failed to connect the 2.6 billion unconnected as of 2025 and miserably failed to *meaningfully* connect hundreds of millions to billions of others. We should remind ourselves what the 'telecoms' market is by revisiting its value chain.

The pre-4G telecoms part of TMT value chain is as depicted in Figure 6.1. As shown, it consists of equipment suppliers like Huawei/Ericsson/ZTE who provide equipment to mobile/cellular operators to build their 2G/3G/4G networks, who in turn provide services like voice calls/SMS to their customers. The wired telecoms subsector value chain has a similar equivalent to Figure 6.1.

Here is the challenge. This is the value chain that has been able to connect the majority of the world's population using 4G (and 5G) as depicted in Figure 6.2, enabling billions to be able to access mobile broadband.

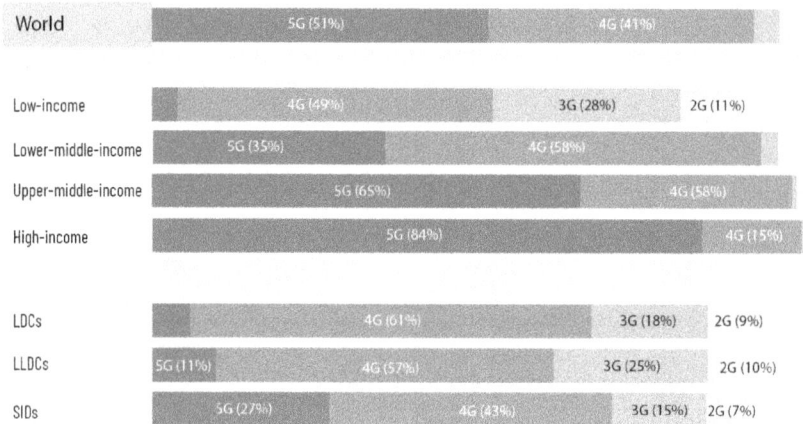

Figure 6.2 – Population coverage by type of network, 2024, Source: adapted from ITU[207]

*[The values for 2G, 3G and 4G networks show the incremental percentage of the population that is not covered by a more advanced technology network (e.g. in 2024, 81 per cent of the 'Low-income'-classified population of the world is covered by at least a 3G or above network (i.e. 4% + 49% + 28%), with 11 percent having only 2G, and 8 per cent having not network coverage at all].*

---

[207] https://www.itu.int/itu-d/reports/statistics/2024/11/10/ff24-mobile-network-coverage/

According to ITU's 2024 data and interactive map[208] (see static subset in Figure 6.2), 4G is now available to 92% of the world's population, leaving a 4G *coverage gap*[209] worldwide of 8%. This is the good news indeed.

The not-so-good news is also demonstrated in Figure 6.2, which amply demonstrates the disparity that the 92% worldwide population coverage with 4G masks. In plain language, the 8% 4G coverage gap in 2024 mostly falls to 'poor' countries – and one really needs 4G ideally to realise 'meaningful' broadband experience, even though 3G may sometimes suffice.

As of end of 2024 as Figure 6.2 [partially] depicts, LICs, LDCs, LLDCs and LMICs have 19%, 15%, 14% and 5% of their populations, respectively, beyond the reach of mobile broadband. So, they are falling short of Target 9.c of Sustainable Development Goal 9: to *"significantly increase access to information and communications technology and strive to provide universal and affordable access to the Internet in least developed countries by 2020."*

Citing from the ITU verbatim:

> "3G or better is now available to 96 per cent of the world population. Bridging the "coverage gap", that is, covering the remaining four per cent that lie beyond the reach of a mobile broadband signal, is proving difficult: since crossing the 90 per cent threshold in 2018, global 3G coverage has increased by only five percentage points. *The largest coverage gap is in Africa, where 14 per cent of the population still does not have access to a mobile broadband network and therefore cannot access the Internet"*[210] (*my emphases*).

---

[208] https://www.itu.int/itu-d/reports/statistics/2024/11/10/ff24-mobile-network-coverage/

[209] This refers to areas of populations that lack of coverage, particularly concentrated in rural and remote areas, especially in regions like Sub-Saharan Africa, which is home to a large percentage (circa 40%) of individuals globally without access to 3G or 4G connectivity (source ITU, *Ibid).*
[210] *Ibid.*

So, the mobile broadband coverage gap disparity falls disproportionately on Africa, specifically Sub-Saharan Africa (SSA) with 14% of the population *not* even being covered in the first place by 3G or 4G.

This is even before the *usage gap* problem sets in. This is because mobile broadband <u>coverage</u> is key to broadband Internet digital inclusion, but it is only part of the solution. This is because there were 3.1 billion people (39% of the global population) living in areas covered by mobile broadband network but *not* using it by the end of 2023[211]. The 2024 number is 3 billion[212]. This is the same usage gap mentioned earlier in this book, e.g. see Section 2.5. As I cover at the introduction of Part II of this book, the reasons for *not* using the Internet are many, complex and varied, including *affordability* challenges of Internet-enabled handsets, high usage data costs, lack of *awareness* of mobile broadband/Internet and its potential benefits, lack of digital skills and confidence to use the Internet, or being prevented from accessing the Internet due to social norms.

Figure 6.3 [from the ITU] shows how even more disproportionately *rural* and acute the digital Internet exclusion problem is. It shows that areas *without* any mobile broadband coverage whatsoever (i.e. where the best available standard is 2G or lower) are *only* found in rural regions[213]. It also shows the proportion of the world's rural population affected by this coverage gap ranges from only 2% in Europe to 25% in Africa. In LDCs, 24% of the rural population are *not* covered, while in LICs it is a whopping 31%. The biggest rural coverage gap afflicts SIDS, where an equally whopping 39% of their populations are not covered by any [mobile] broadband Internet access at all.

However, even Figure 6.3 does *not* tell the full story of those who are *not* covered by any mobile signals whatsoever because it is about those without access to 3G. According to the same ITU's Facts and Figures 2024 report[214], 2% of the world's population is *not* covered by 2G basic voice signals as of 2024 <u>at all</u>. This means

---

[211] https://www.gsma.com/r/wp-content/uploads/2024/10/The-State-of-Mobile-Internet-Connectivity-Report-Key-Findings-2024.pdf

[212] https://www.gsma.com/solutions-and-impact/connectivity-for-good/mobile-economy/wp-content/uploads/2024/02/260224-The-Mobile-Economy-2024.pdf

[213] Personally, I do not even consider 3G to constitute enabling broadband because it hardly offers 5 Mb/s to subscribers.

[214] https://www.gsma.com/solutions-and-impact/connectivity-for-good/mobile-economy/wp-content/uploads/2024/02/260224-The-Mobile-Economy-2024.pdf

that – usage gap aside – circa *175 million people in 2024* are *not* even living under a mobile signal to enable them to effect basic 2G voice calls.

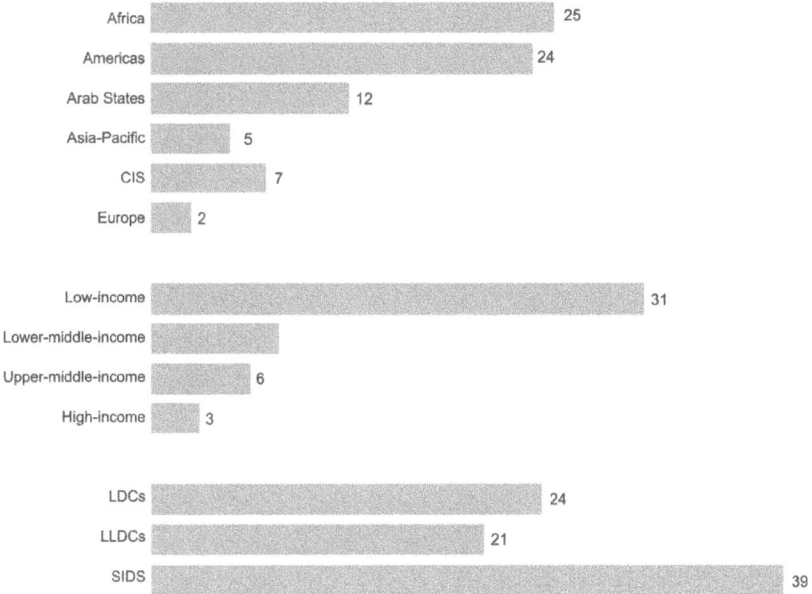

Figure 6.3 – Percentage of the *rural* population *without access* to a 3G mobile network or higher, 2024 - Source: adapted from ITU[215]

Given the above narrative on mobile broadband (and mobile voice), there should be no prizes for guessing where the totally unconnected reside. So, where does the 2% of the world's population *not* covered by 2G basic voice signals live? They primarily reside in rural and remote areas, particularly in Sub-Saharan Africa (SSA) and some parts of Asia and Latin America. An example of this can be seen in Democratic Republic of the Congo in SSA, which had one of the highest coverage gaps: 25% do *not* have any mobile coverage (including 2G)[216] and 46% of the population do not have mobile broadband coverage.

---

[215] https://www.itu.int/itu-d/reports/statistics/2024/11/10/ff24-mobile-network-coverage/
[216] *Ibid.*

### 6.1.1 Telecom Market 'Failure': 1.3 Billion Rural Users *Not* Using 2G Voice Access

The telecoms market has 'failed' key [rural] populations – with a huge usage gap of 1.3 billion [minimum] *not* using basic 2G voice services today in 2025. The title of this section is the clear conclusion from the preceding narrative. It is uncomfortable, but inescapable. And – usage gap aside – this pertains to basic 2G services to enable circa 175 million people worldwide to be able to make a basic telephone call in 2024/25. I write 'failure' – i.e. in quotes – because it is arguably unfair to put all or most of the blame at the foot of the telecoms industry, thanks to usage gap issues like poverty, unaffordable phones, excise duties on phones, etc. However, the 1.3 billion or the 175 million have been 'failed' nonetheless!

2G voice services do *not* present the same usage gap challenges that using the mobile Internet presents. The latter requires more expensive Internet-enabled devices – but more crucially, it assumes literacy, some degree of education, awareness of the benefits of going online, digital skills, etc. Making a basic phone call does not require these, yet 175 million people in 2024/25 could not make one. These people are suffering from a 2G coverage gap issue. [Note that 2G does not allow users to have any meaningful Internet experience and service as it really does not support mobile broadband].

Adding to this circa 175 million coverage gap number is the 2G *usage gap* number, i.e. where there is at least basic 2G coverage, yet it is not being used for whatever reasons. As of 2024, mobile phone ownership was at 80%[217] (for those over 10 years old). This is good – but it starts telling us a story about what percentage of the population do *not* even make basic voice calls regularly. There appears to be an 80%-90% mobile voice usage rate worldwide as of 2024, which means that there is a 10%-20% *usage rate* gap, even for basic 2G voice alone for whatever reasons. I derive and/or infer this from the respected sources of the GSMA and the ITU.

- Digging through ITU's data, you realise that in Africa, for example, 18% of the *rural population* lacks any mobile network coverage, and another 11% has only 2G coverage as of 2024[218].

---

[217] *Ibid.*
[218] https://www.itu.int/itu-d/reports/statistics/2024/11/10/ff24-mobile-network-coverage/

- The GSMA's 2023 State of Mobile Internet Connectivity report[219] reports on some fascinating GSMA Consumer Survey coverage and usage gap analyses across the globe, focusing on 12 LMICs[220]. Though the analyses in the report focused mainly on mobile Internet connectivity, it does provide some interesting insights as regards basic mobile connectivity for just 2G voice. For example, *inter alia*, the report concluded that the

> "majority of the usage gap comprises those *without a mobile of any type* … and more than 600 million people (8% of the global population) who have a basic or feature phone, are using mobile services but are *not* connected to mobile internet. The remaining two thirds of the usage gap *do not own a mobile*"[221] (*my emphases*).

This "remaining two thirds of the usage gap do *not* own a mobile" is circa **2 billion people** at the end of 2023[222], i.e. 20% of humanity. Many of these 2 billion likely live in households with others that have a mobile/cellular phone – but the key insight is that even a basic commodity service like voice is not *regularly* being used in 2025 by a significant proportion of humanity, most probably a billion people or more. The GSMA's 2024 State of Mobile Internet Connectivity report[223] largely confirms these findings.

---

[219] "The State of Mobile Internet Connectivity Report." https://www.gsma.com/r/wp-content/uploads/2023/10/The-State-of-Mobile-Internet-Connectivity-Report-2023.pdf

[220] Bangladesh, Egypt, Ethiopia, Ghana, Guatemala, India, Indonesia, Kenya, Mexico, Nigeria, Pakistan and Senegal.

[221] https://www.gsma.com/r/wp-content/uploads/2023/10/The-State-of-Mobile-Internet-Connectivity-Report-2023.pdf, p.17.

[222] The math I have derived from the GSMA 2023 Report is as follows: Start with 3 billion people who are covered by mobile broadband networks but are *not* using mobile internet. 1/3 of this number were found to have a primary mobile device which they, at least, use for mobile voice/SMS. This third includes 350 million smartphone owners using mobile services but *not* mobile internet, and more than 600 million people who have a basic or feature phone, are using mobile services but are not connected to mobile internet. The remaining two thirds of the usage gap do *not* own a mobile/cellular phone at all. So, this 2/3rds of 3 billion equates to 2 billion people.

[223] https://www.gsma.com/r/wp-content/uploads/2024/10/The-State-of-Mobile-Internet-Connectivity-Report-2024.pdf

- The ITU's Facts and Figures 2024 report[224] is more nuanced on the 2 billion number above, as it found that four out of five people *over 10 years old* own a mobile/cellular phone – ergo 20% do *not*. It however says major differences existed between countries, as more than 95 per cent of people over 10 years old own a mobile phone in HICs, compared to only 56 per cent in LICs. And according to the World Bank[225], approximately 26% of the world's **8.09 billion** population is under the age of 15. So, to estimate the population excluding those under 10 years old, I used a simple approach:

  - Population under 15 years old: 26% (Source: World Bank, 2024)
  - Population under 10 years old: circa roughly 20% (assuming a uniform distribution within the 0-15 age group)
  - So, population *over* 10 years old would be circa 8.09 billion × (1−0.20) = 6.47 billion
  - 20 percent of this number do *not* own a mobile phone (since 80 percent of this number own a mobile/cellular phone) i.e., 6.47 billion x 0.2 = **1.294 billion.**

Therefore, 1.294 billion people at the beginning of 2024 <u>did not *own*</u> a mobile/cellular phone to effect even the most basic 2G voice call.

This leads us to this simple summary table shown in Table 6.1 of those who cannot even make a basic 2G voice call as at the beginning of 2024 – drawing from the GSMA and ITU numbers above.

| Coverage Gap Worldwide (GSMA) | 175 million |
|---|---|
| Usage Gap (ITU) – *derived above* | 1,294 million |
| Usage Gap (GSMA) - *inferred above* | 2,000 million |
| **Total (Range)** – assuming coverage and usage gap *are* <u>mutually exclusive</u> | **1,469 – 2,175 million** |
| **Total (Range)** assuming coverage and usage gap *are not* <u>mutually exclusive</u> | **1,294 – 2,000 million** |

Table 6.1 – Estimate of Basic [2G] *Voice Usage Gap*

[224] https://www.itu.int/itu-d/reports/statistics/2024/11/10/ff24-mobile-phone-ownership/
[225] https://data.worldbank.org/indicator/sp.pop.0014.to

My warning in this section specifically pertains to both the coverage and usage gaps concerns that exist to date, even for basic 2G voice services – as shown in Table 6.1.

*Individually*, each gap [coverage or usage] is already an extremely hard issue for the telecoms market and its value chain (shown in Figure 6.1) to address. *Together*, they self-reinforce each other making the problem very intractable and difficult to unpick.

Table 6.1 informs us that – as of the beginning of 2024 – some number between 1.449 billion to 2.175 billion people globally could not *regularly* make a basic 2G voice call. 175 million would have to travel to a place where there is even basic 2G coverage (the coverage gap problem) and somewhere, conservatively, in the range 1,294 – 2,000 million people (above 10 years of age) could *not* effect a basic call regularly (the usage gap problem). However, they may have access in their homes to others who own a device.

In very plain English language, the telecoms market has failed – and is failing – to connect more than a billion (up to 2 billion) people globally for them to affect a most basic voice commodity service through a combination of *coverage gaps* and *usage gaps* market failures. This is the grim reality in 2024/25, thirty-three (33) years after the first 2G network and GSM call was made in Finland[226].

### 6.1.2  2 Billion 'Connected' Users Lack Meaningful Access

2 billion of those 'connected' are *not* meaningfully connected. Given the message of Table 6.1 earlier above, I ask rhetorically whether the market failure for connecting the global population online [meaningfully] would be easier or harder – given the more complex usage gap challenges like illiteracy, lack of education, lack of awareness, poverty and more? What really are the chances of connecting the tail 2.6 billion people online when possibly 2 billion of them (or more) either do *not* have a mobile/cellular device and/or do *not* even regularly make basic 2G voice calls?

---

[226] https://www.gsma.com/about-us/who-we-are/our-history/

I think the answers to these questions are obvious. GSMA's finding of 3 billion people (37% of the global population) living in areas covered by mobile broadband network but *not* using it, as of the end of 2024[227] is very telling. The ITU also notes that:

> "in Africa, although 66 per cent of the population own a mobile phone, only 38 per cent are online, a difference of 29 percentage points. Still, the gap is shrinking in all regions, *as growth in Internet use continues to outpace growth in mobile phone ownership*"[228] (*my emphasis*).

I am intrigued by this ITU finding: "as growth in Internet use continues to outpace growth in mobile phone ownership". This is probably true, but there appears to be a 'floor' that is difficult to go below, i.e. circa possibly 2 billion (or more) people who – as we see earlier - neither do *not* have a mobile/cellular device nor even regularly make basic 2G voice calls.

GSMA found in 2022 that the *Mobile Internet's 'Usage Gap' is Almost Eight Times the Size of the 'Coverage Gap'*[229]. This fact, the likely 2 billion+ floor above, in addition to much more difficult usage gap challenges like illiteracy – collectively support my proposition that the telecoms market will fail the majority of the 2.6 billion unconnected even more.

To put my proposition in more *stark* terms, consider Table 6.2 which uses the data shown in Figure 1.1, i.e. the *head* 2.7 billion rich and always online, the *mid-tail* 2.7 billion who can get online as needed, albeit many of them not *meaningfully connected*, and the *tail* 2.6 billion poor who are offline in 2025". In Table 6.2, I show three assumptions on what proportion of the mid-tail 2.7 billion [mostly] are truly meaningfully connected online: a low point assumption of 10%, a mid-point of 25% and a high point of 35%. This is because – as I cover in Chapter 1 (Section 1.2) – it is safe to deduce that a substantial portion of those online are *not* "meaningfully" connected, due to factors like electricity/power, unreliable access, high usage costs, affordability issues, and lack of sufficient digital skills.

---

[227] https://www.gsma.com/solutions-and-impact/connectivity-for-good/mobile-economy/wp-content/uploads/2024/02/260224-The-Mobile-Economy-2024.pdf
[228] https://www.itu.int/itu-d/reports/statistics/2024/11/10/ff24-mobile-phone-ownership/
[229] https://www.gsma.com/newsroom/press-release/mobile-internets-usage-gap-is-almost-eight-times-the-size-of-the-coverage-gap-gsma-research-reveals/

| | | |
|---|---|---|
| Tail Poor who are offline in 2024/25 (ITU) | 2,600 million | 2,600 million |
| Mid-Tail **not** meaningfully 'connected' 2024/25: assume 90% (i.e., only 10% meaningfully connected) | 2,700 million x 0.9 = 2,430 million | |
| Mid-Tail **not** meaningfully 'connected' 2024/25: assume 75% (i.e., only 25% meaningfully connected[230]) | 2,700 million x 0.75 = 2,025 million | 2,025 million |
| Mid-Tail **not** meaningfully 'connected' 2024/25: assume 65% (i.e., only 35% meaningfully connected) | 2,700 million x 0.65 = 1,755 million | |
| **Total (Mid-point of Range)** | | **4,625 million** |
| **Total (Range)** | | **4,355 - 5,030 million** |

Table 6.2 – Estimates of Unconnected and *not* Meaningfully-connected Internet Usage Gap

Just in case the reader believes that my *low-medium-high meaningful-connectivity assumptions* are too pessimistic at 10%-25%-35% respectively, I present a more authoritative basis for them in Table 6.3. The Alliance for Affordable Internet (A4AI) used their well-researched *Meaningful Connectivity framework[231]* and applied it to nine LMIC economies (Colombia, Ghana, India, Indonesia, Kenya, Mozambique, Nigeria, Rwanda, and South Africa), using *real* mobile phone surveys to estimate the number of people with meaningful connectivity in each of these LMIC countries[232].

---

[230] Justified by Table 6.3 and its following narrative.
[231] The framework focuses on four pillars: 4G-like speeds, smartphone ownership, daily use, and unlimited access at a regular location, like home, work, or a place of study.
[232] https://a4ai.org/report/advancing-meaningful-connectivity-towards-active-and-participatory-digital-societies/

| COUNTRY | NATIONAL | URBAN | RURAL | Men | WOMEN |
|---|---|---|---|---|---|
| Colombia | 26.20% | 30.50% | 7.60% | 33.80% | 19.20% |
| Ghana | 6.50% | 9.00% | 2.80% | 8.30% | 4.80% |
| India | 6.80% | 9.00% | 5.30% | 9.80% | 3.30% |
| Indonesia | 12.70% | 15.30% | 9.10% | 12.80% | 10.40% |
| Kenya | 10.90% | 20.70% | 6.50% | 14.30% | 7.70% |
| Mozambique | 3.60% | 6.70% | 1.50% | 4.20% | 2.70% |
| Nigeria | 12.10% | 16.40% | 6.60% | 15.50% | 7.20% |
| Rwanda | 0.60% | 1.90% | 0.30% | 0.50% | 0.20% |
| South Africa | 12.80% | 15.90% | 5.70% | 16.40% | 12.10% |

Table 6.3 – Estimates of meaningful connectivity, by country, geography, and gender
Source: Adapted from A4AI (2022)

The reader can see from Table 6.3 that the national meaningful connectivity proportions ranged from 0.6% in Rwanda to 26.2% in Colombia.

Based on the data presented in Table 6.3, the low, medium, and high meaningful-connectivity assumptions are set at 10%, 25%, and 35%, respectively – as justified following.

- 10% is a cautiously chosen *low estimate* for meaningful connectivity across LMCs, given that India with a population of 1.4 billion was assessed at 6.8% meaningful connectivity.
- 25% is arguably a too-optimistic *'mid-point' estimate assumption*, given that, nationally, only Colombia exceeded this number in the 2021 research. However, I have still chosen to 'fend off' accusations of being too pessimistic by using this number as my mid-point.
- 35% is arguably yet another too-optimistic best case of meaningful connectivity across the LMIC countries given that no urban setting in Table 6.3 even shows any number that is above 30.4%.

So, returning to Table 6.2, *the estimates of the unconnected and the not meaningfully connected Internet Usage Gap* across the world suggests a mid-point of 4,625 million, and a range between 4,355 - 5,030 million. The size of the challenge is therefore clearer, I hope – 2.6 billion completely offline, and a midpoint of another 2 billion not meaningfully connected.

### 6.1.3 The Telecoms Market Will Fail Most of the Tail 2.6 Billion Not Online *Even* More

Therefore, it seems very logical to conclude that telecoms market – that has already failed key [rural] populations, even for basic 2G voice services (cf. Section 6.1.1) – will almost certainly fail the majority of the 2.6 billion unconnected even more.

In summary, I conclude this Section 6.1 by positing that it is unfair to *mostly* put the blame [of the 2.6 billion offline and the 2 billion not meaningfully connected] at the foot of the telecoms market. This is because – what are they really supposed to do about the usage gap problem, wherein 3 billion (at the end of 2024) are not using the [mobile] Internet even though they live under mobile broadband signals? This is why I title this entire section [Section 6.1]: *Warning: beware of waiting for solutions to the 2.6 billion from the telecoms market.* The problem is much more intractable, particularly for LMICs and LICs who dominate the unconnected and the not-so-meaningfully connected.

## 6.2 Regulatory and Policy Interventions are Therefore Very Necessary

So, what are we supposed to do if the telecoms market is failing, has failed and will continue to fail to online-connect most of the 2.6 billion and meaningfully connect another 2 billion more? I conclude in the previous section by positing that it is unfair to *mostly* put the blame on the telecoms market. Broadly, the market depends on *competition* amongst telecommunication licensees (competing fairly against one another) to provide as much coverage of people as possible. However, competition and the market can only go as far as the *rational profit-maximising market players perceive is beneficial to them*, and their shareholders. Chapter 2 provides examples from the Papua New Guinea (PNG) and Malawi telecom markets – elaborating on why the telecom markets in these two countries have so failed to connect significant proportions of the citizens of these two great countries. So, markets naturally fail for many consumers and citizens in society, including the sort of usage gap reasons covered in this chapter – the so-called *market failure* theory.

The telecoms example shown in Figure 6.4 illustrates such a market Failure theory. It depicts that regulation is needed where there is (or would be) a market failure, i.e., where competition fails.

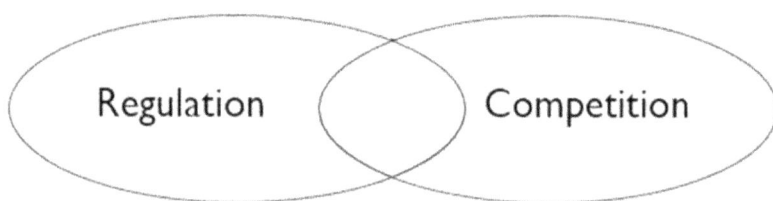

Figure 6.4 – Regulation vs. Competition 'Toolboxes'

Figure 6.4 shows that – in order to seek to achieve positive outcomes for consumers and citizens that do <u>not</u> arise naturally from market *competition* [on the right of Figure 6.4], i.e. a market failure – *regulation* [on the left of Figure 6.4] becomes necessary to bring net benefits.

Put another way, Figure 6.4 shows that if competition is working fine for all consumers and citizens, there is *no need* for the intervention of regulation (Nwana, 2024). However, if competition is failing a large proportion of the citizens, then the intervention of regulation becomes very necessary.

Hailing from a LIC/LMIC country myself [Cameroon], I am confident to assert that online-connecting 2.6 billion and meaningfully connecting another 2 billion more would require more *regulatory toolbox instruments*, than continuing to rely on competition and the market (cf. Figure 6.4). This must be understood by all those who I address this book to.

Many-a-time, I observe closely that the leaders of LIC, LMIC and/or LDC economies appear to abdicate all online connectivity to their telecoms markets.

Given the significant challenges I elaborate on in Section 6.1, I assert that regulatory intervention is therefore the key consideration going forward.

## 6.3 Caution 2: Beware of the Hype and Promises of [New] Sexy Technology Solutions

I initially titled this chapter – *No more market-led telecoms solutions to connect the tail 2.6 billion* before I changed it to what it is now. This is my second caution of the chapter: LMIC/LIC/LDC leaders should beware of the hypes and promises of all the 'sexy' technology solutions that promise to connect most of the tail 2.6 billion unconnected. Telecoms history to date has comprehensively showed that lasting connectivity solutions in LMIC/LIC/LDC regions *must also be pragmatic and need-centred*, not just market-driven – as is clearly illustrated with the M-Pesa mobile financial services (MFS) solution in Kenya – that I describe in Chapter 5.

The reader may, or would likely have, heard of technologies like 5G, perhaps even 6G, LEOs, MEOs, HAPs, HIBs, TVWS, optical fibre, or projects like Starlink, Oneweb, Project Kuipers, etc. I could go on. I acknowledge that I have advocated for some of these technologies in low-income, lower-middle-income, and low-income-to-middle-income countries. The reality is that the digital divide continues to persist and even get worse. Permit me to revisit the promises (and hypes) of these technologies, and the reality vis-à-vis these technologies addressing the digital divide.

i.   *5G (fifth generation) completely ignores the 2.6 billion offline:* 5G promised to deliver significantly *higher data speeds* compared to 4G, reaching up to 10 Gbps. This hype revolved/revolves around quicker video and audio streaming, faster downloads, and more seamless browsing experiences. It promised/promises *low latency* (i.e., reducing online lag to as little as 1 millisecond), *improving reliabili*ty (even in densely populated areas) by reducing instances of dropped calls and buffering. This is/was expected to revolutionise applications requiring instant responses, real-time applications like gaming or remote surgeries. 5G promises/promised massive *machine-to-machine communications*, i.e., the number of connected devices per square kilometre, including smart homes, autonomous cars, and connected cities.

The reality of 5G to date vis-à-vis the digital divide is already clear to observe. A friend and colleague of mine authored a celebrated book which he titled *The 5G Myth: When Vision Decouples from Reality* (Webb, 2018). His analyses and criticisms have proved mostly right: 5G speeds are just incremental to 4G, and true 5G connectivity has remained limited and niche: even in the US, only 18.8% of AT&T subscribers, 30.1% of T-Mobile's, and 9.5% of Verizon's connect to 5G as of 2023/24[233]. The costs for implementing 5G are also extremely high. As of early 2025, my personal iPhone 15 usually latches onto 5G here in the UK where I live, but I frankly prefer to be on 4G – as I find it both more seamless and dependable.

Bottom line, the reality is that 5G offers extraordinarily little to connect the 2.6 billion offline. It clearly offers more capacity and speeds to the head 2.7 billion who are already ultra-connected (cf. Figure 1.1) in countries like South Korea, UK, Europe/EU, Japan, Australia and the USA.

Why LMIC, LIC and LDC leaders, ICT ministers and regulators keep falling for, and keep advancing, the 5G hype truly befuddles me. If you do not believe me, believe Figure 6.5 from Ericsson. Note the minimal impact 5G would have by 2030 in Sub-Saharan Africa (circa 35%), and Southeast Asia & Oceania (circa 51%), according to Ericsson's own projections. You could argue that projections for other regions look positive. The Middle East and North Africa, Latin America, India, Nepal, and Bhutan are regions where many individuals remain unconnected. As a result, they will likely *not be included in these projected 5G figures from Ericsson.* These projections, which I find optimistic, would largely represent upgrades from existing 3G/4G subscriptions – not new connections from the 2.6 billion unconnected.

---

[233] https://www.statista.com/topics/3447/5g/#topicOverview

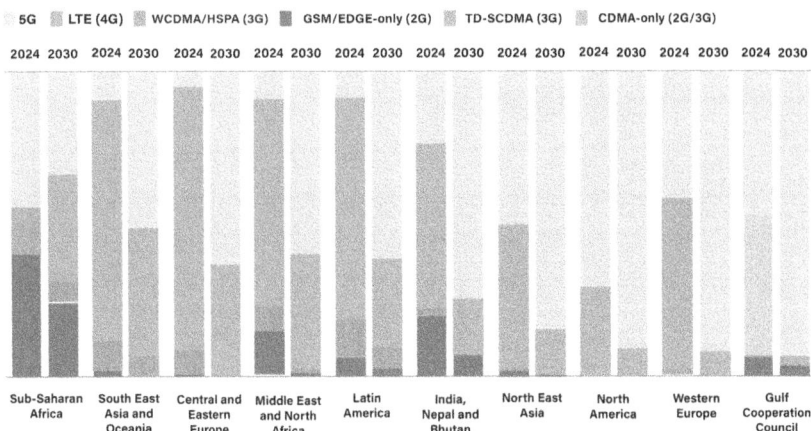

5G    LTE (4G)    WCDMA/HSPA (3G)    GSM/EDGE-only (2G)    TD-SCDMA (3G)    CDMA-only (2G/3G)

Figure 6.5 – Mobile Subscriptions by Region and by Technology
Source (Adapted from Ericsson Mobility Report[234], November 2024)

ii. ***Forget 6G (sixth generation) for connecting the 2.6 billion unconnected:***
the coming 6G promises more of the same similar 5G hypes: "5G on
steroids", unprecedented advancements in connectivity, etc. 6G will be
similarly tangential to connecting the 2.6 billion offline as 5G has been. In
his 6G manifesto book, Webb (2024) argues that 6G should address, *inter
alia*, the coverage issue and seamless integration with Wi-Fi – i.e., prioritise
connecting more of those 2.6 billion who are not online today. For example,
he fears the story will be remarkably like 5G, which hardly considered
coverage.

iii. ***Satellite solutions have long offered significant potential for connecting
the billions who are <u>not</u> online; however, they have often failed to meet
expectations***: Satellite access encompasses connectivity via Low Earth
Orbit (LEO), Medium Earth Orbit (MEO), or Geostationary Earth Orbit
(GEO) satellites. The biggest advantage of satellite is its widespread
footprint: three equally spaced GEO satellites 35,700 km above the equator
can cover almost the entire surface of the earth. There have been so many

---

[234] https://www.ericsson.com/4adb7e/assets/local/reports-papers/mobility-
report/documents/2024/ericsson-mobility-report-november-2024.pdf

advancements in the satellite sector through new LEO constellation players like Amazon Project Kuiper, OneWeb and Telesat, SpaceX StarLink[235] whose activities have brought launch costs down circa 75%.

Starlink is arguably shifting – at the edges – the competitive landscape for rural connectivity, prompting terrestrial telecom companies to look for opportunities in previously underserved areas. I have seen this early trend in Africa as of early 2025 with several tie-ups or ventures: MTN exploring LEO satellite partnerships for Rural Connectivity[236]; Orange and Vodacom announced a deal in Democratic Republic of Congo for network deployment in rural areas[237]; Safaricom [Kenya] announced a partnership with local satellite operator ESD Kenya[238]; etc. There appears to be some Starlink-instigated, newfound desire to boost connectivity across the footprint of several African countries via LEOs. I commend this trend, but I am wary it will be short-lived.

This is because these satellite technology solutions – no matter the type (i.e., GEOs, MEO, LEOs) – have always appeared like a silver bullet for connecting those billions offline, but the reality has often fallen well short. The reasons and several and varied:

a. *Affordability*: the cost of satellite Internet services and the required hardware are typically prohibitively expensive for most in low-income and remote rural areas in LICs and LMICs.

b. *High initial and ongoing costs*: installing and maintaining satellite infrastructure requires substantial investments. The cost of launching satellites and setting up ground stations often makes it less viable compared to other technologies. LEO satellites have shorter life spans of circa 7 to 10 years. GEOs have an average life span of 15 years, whilst MEOs last between 10-15 years. This means that satellite solutions tend to be more expensive – on an ongoing basis – in terms

---

[235] https://www.economist.com/business/2021/05/13/spacex-a-tesla-for-the-skies
[236] https://businessmetricsng.com/mtn-explores-satellite-partnerships-for-rural-connectivity/
[237] https://newsroom.orange.com/orange-and-vodacom-create-a-joint-venture-to-expand-network-coverage-in-rural-areas-in-the-drc/
[238] https://broadcastmediaafrica.com/2025/02/12/report-on-starlinks-impact-on-telecoms-companies-in-africa/

of the number of satellites required and how often they need to be replaced, particularly so with LEOs.

c.  ***Physics still matters with propagation from satellites and FSOC[239]-based solutions***: this is a basic truism that most will miss from the hype surrounding satellites, FSOC-based solutions like Taara, and even with LEOs like Starlink. Basic physics tells us that the power received in each area decreases with the square of the distance from the source, i.e. as the distance doubles, the signal becomes 4 times weaker. Put another way, to double the coverage radius requires a 4-fold increase in power, and doubling the coverage area requires a 2-fold increase in power. coverage. This is called the *Inverse Square [Power] Law.*

A terrestrial mobile signal from a terrestrial 2G/4G base-station typically covers a radius of 0.5 to 4km. In contrast, a LEO Starlink satellite signal travels a great distance to reach Earth from an altitude of 550 km (many LEOs are higher than this[240]).

So, let us consider a scenario [Scenario 1] of a terrestrial 5G 3,500MHz signal being transmitted downlink from a base-station reaching a 5G device 2.5km away. Consider another scenario [Scenario 2] of a LEO satellite signal on the same 3,500MHz frequency from an altitude of 550km. 550km divided by 2.5km is 220 times the distance away. So, with everything else being equal, the *Inverse Square Law* dictates that the signal from the satellite will be $(220)^2$ times weaker to the 5G device – and this assumes the device is outdoors with reasonable clear line of sight to the satellite – compared to from the terrestrial base-station. This means a 48,400 times weaker signal for the same frequency from the satellite in comparison to a terrestrial signal. If the device is indoors, the challenge increases markedly again, due to increased signal attenuation when passing through walls etc. This is significant given that 80% or more of our mobile/cellular broadband consumptions happens indoors according to Ericsson[241].

---

[239] See later for FSOC
[240] The altitude range of the LEO satellites are typically 300 – 1200 km.
[241] https://www.ericsson.com/en/reports-and-papers/mobility-report/articles/indoor-outdoor

So, satellite D2D physics is most challenging, and therefore terrestrial networks will always win, or be preferred. And this is just for the downlink from the satellite to the device. The uplink from a 5G device to a satellite is significantly challenging due to mobile devices' low transmission power at circa 20–30 dBm[242]. This makes it difficult for uplink signals to reach satellites in LEO orbits 550 km above, especially outdoors, and even more so indoors.

d. *Latency Issues:* due to the significant distances that the information must travel to and from satellites, time lag or latency can also be a significant issue. Even though LEO satellites[243] have improved latency, it remains an issue. High latency creates low quality of experience issues with real-time applications like video conferencing and online gaming.

e. *Dependence on Weather Conditions and Preference for Fibre and Submarine Cables*: satellite signals, due to fundamental physical principles, are prone to interference from adverse weather conditions. Heavy rains, geomagnetic storms, and even heavy clouds disrupt satellite connectivity, making it unreliable compared to undersea and terrestrial alternatives. For such reasons as costs, latency and weather challenges, the best information suggests that satellites carry less than 1% of the world's Internet/online traffic[244], and we rely on submarine cables almost for the rest of the 99% – because they are both faster and cheaper than satellites.

f. **Control of Satellites**: Billionaire Elon Musk allegedly plans to launch up to 42,000 low-orbit satellites as part of his SpaceX programme[245].

---

[242] Apologies to the non-technical reader about this technical measure. 20dBm is 0.1 Watt of power whilst 30dBm is 1 Watt of power. For context, some high-power terrestrial TV broadcast transmitters put out power at 20 kilowatts!

[243] LEO satellites, usually operating at altitudes of 300-2,000 km (186-1243 miles) above the Earth's surface, provide for much better latencies than GEO or MEO satellites, which are placed in much higher orbits.

[244] https://researchoutreach.org/articles/satellite-internet-technology-double-edged-sword/

[245] As of February 2025, Starlink had already launched over 7,000 satellites providing connectivity to over five million customers in more than 100 countries and aims for 12,000 more from 2025 on. Source: https://www.telecomreview.com/articles/reports-and-coverage/8756-the-challenges-and-limitations-of-leo-satellites/

As of January 2025, Starlink had a network of almost 7,000 satellites operating, including 300+ (2nd generation or Gen2) satellites which have Direct-to-Cell (D2C) capabilities. SpaceX allegedly plans to deploy an additional 30,000 Gen2 satellites with D2C capabilities, but no official specified timeline for this deployment has been announced. Similarly, Amazon's founder Jeff Bezos plans to launch 3,236 satellites through his $10 billion Kuiper project. However, as of January 2025, Amazon's Project Kuiper had not yet launched any operational LEO satellites, with similar D2C capabilities to Starlink's. Satellites have also been launched by so-called 'problematic' states like Iran, China, Russia and North Korea.

So, many countries, LICs and LMICs included, are justifiably nervous of relying for their Internet connectivity on such powerful individual billionaires like Musk or Bezos, or on other 'problematic' states (Naim, 2023). For instance, in early March 2025, Mr. Elon Musk indicated that Ukraine's utilisation of the Starlink LEO satellite system, specifically within the context of the Ukraine-Russia conflict, is critical. He stated,

> "*My Starlink system serves as the backbone of the Ukrainian army.*" Their entire front line would collapse *if I turned it off*"[246] (*my emphases*).

Indeed, the June 2025 fallout between Elon Musk and Donald Trump escalated dramatically, with Elon Musk even threatening to decommission SpaceX's Dragon spacecraft, which is crucial for NASA's missions to the International Space Station (ISS)[247]. This came after Trump threatened to cut government contracts with Musk's companies, including SpaceX and Starlink[248]. Musk initially announced that SpaceX would begin decommissioning Dragon immediately but later walked back on the threat after being urged to

---

[246] https://www.bbc.com/news/articles/cy87vg38dnpo &
https://www.telegraph.co.uk/world-news/2025/03/09/musk-ukranian-front-line-collapse-starlink-turned-off/
[247] https://www.pbs.org/newshour/politics/musk-backs-off-threat-to-decommission-dragon-space-capsule-amid-feud-with-trump
[248] https://www.reuters.com/world/us/trump-blames-musks-criticism-decision-cut-ev-tax-credits-2025-06-05/

reconsider. Given that SpaceX holds around $22 billion in government contracts, including key NASA and Pentagon projects, this dispute could have serious implications for the U.S. Space Program.

It is frightening to me that a billionaire owner could possibly switch off Starlink to Ukraine and threaten the U.S. Space Program, and countries are justified to be wary of such utterances from satellite-controlling billionaires. Indeed, Starlink is neither available nor appears to have any plans to launch in China, Iran, Russia, Afghanistan, Syria, Cuba, North Korea, and Belarus. This is because these countries are considered political and/or economic enemies of the USA. Starlink has also struggled – and is still struggling as of March 2025 – with approvals in South Africa, India, Pakistan, Bangladesh, Vietnam, and many African countries. Part of the reason is because Starlink wants to go directly to customers for its residential service, bypassing any partnerships with local Internet Service Providers (ISPs). Sovereignty concerns also loom large with many countries. This would allow others like China's SpaceSail to enter these markets faster. SpaceSail – the so-Chinese "Starlink" – is launching in over 30 countries. It already launched its first 18 satellites for LEO connectivity on 6 August 2024 and by January 2025 already had 72 satellites in orbit[249]. "G60 Starlink" aims to offer worldwide Internet coverage in the coming years. So, the geopolitics of satellites is alive and well.

g. *Regulatory challenges (spectrum regulation, cybersecurity, data sovereignty, etc.)*: obtaining the necessary authorisations and licences to install and operate satellite equipment typically takes time due to various bureaucratic policies, competitive factors, and regulatory processes, which differ from country to country based on their specific requirements. Satellite D2C/D2D relies on spectrum that local regulators have not officially licensed to LEO players, thereby creating tensions with governments and local telecom providers, e.g. for 3,500MHz spectrum. Cybersecurity (transparency) and data sovereignty are increasingly concerns too with many countries. These checks typically slow down satellite deployment.

---

[249] https://www.reuters.com/technology/musks-starlink-races-with-chinese-rivals-dominate-satellite-internet-2025-02-24/

h.  *Technical challenges*: satellite solutions need to be precisely installed, and the alignment of the satellite dishes can be challenging. This includes securing permissions to install on rooftops or high locations for optimal signal reception.

*My summary on satellites*: therefore, here is my overall take on the prospects of satellite solutions truly addressing the digital divide. Mobile/cellular networks have *not* replaced fibre or other fixed networks. So, satellite networks – no matter their hue as in GEOs, MEOs or LEOs - do *not* have the capacity (and physics) to replace mobile/cellular networks in most areas. The reality is that terrestrial mobile networks will remain the primary form of connectivity for most unconnected peoples. I acknowledge that some major MNOs like Telstra [Australia] have announced collaboration with LEOs like SpaceX's Starlink to deliver direct-to-device (D2D)/direct-to-handset outdoor text message capability to parts of the country that are beyond the reach Telstra's terrestrial network[250]. Ditto T-Mobile's collaboration with SpaceX's Starlink in the USA[251]. Telstra notes that direct-to-handset has the potential to be a game changer from a safety perspective for people living and travelling in remote Australia[252]. Recall the challenges due to physics that I note above. These D2D/D2C benefits are clearly complementary, but niche.

iv.  ***Do not wait on satellites to bridge the bulk of the 2.6 billion unconnected:*** in summary, I am very wary of satellite operators who have a very long history of promising to bridge the digital divide in LDCs, LICs and LMICs, but delivery has often fallen well short of their promises. So, there are many LMICs and LICs who are increasingly wary of these promises with good reasons too, and even warier of the billionaires who own or control these LEO constellations. So, the reality is that satellite will/would continue to remain very niche to connecting the 2.6 billion not online without significant [Government] subsidies. For example, the Government of

---

[250] https://www.telstra.com.au/aboutus/media/media-releases/telstra-brings-spacex-starlink-satellite-to-mobile
[251] https://www.t-mobile.com/coverage/satellite-phone-service
[252] *Ibid.*

Canada is partnering with Telesat to offer a connectivity model that benefits rural customers and telecom operators – in which telecom operators have access to a dedicated pool of Telesat's LEO capacity in order to serve rural households, subsidised through Government investment, hence Telesat can proceed to offer reduced rates to telecom operators[253].

v.  **HAPs and HIBs:** in 2018, I published a comprehensive "High Altitude Platforms (HAPS) for Affordable Broadband Connectivity" whitepaper[254]. The ITU defines HAPS in the Radio Regulations "as a station located on an <u>object</u> at an altitude of <u>20-50 km</u> and at a specified, nominal, fixed point relative to the Earth, and is subject to [Article] No. 4.23"[255]. HAPS solar aircraft are frankly not new, and they have been pioneered for decades – previous attempts were arguably ahead of their time. Today, unlike in the 1990s, HAPS are becoming reality – slowly – as advancements primarily in battery technology[256] (lithium batteries), high efficiency solar panels, lightweight materials, autonomous aircraft avionics and microwave payloads have combined to make HAPS solar aircrafts being able to provide communications backhaul service realisable in the medium term. The HAPS concept is one of an unpiloted High-Altitude, Long Endurance (HALE) platform which relays telecommunications from Gateway stations and user terminals on the ground. HAPS solar aircrafts are not quite HALE-yet in 2024/25, but have already demonstrated continuous flights over multiple days, which are being extended to several weeks and months, if not up to a year. HAPS are still subject to ongoing tests.
Facebook/Meta's Aquila Drone[257] and Google's Loon[258] projects were example HAPs projects which these companies have sadly and quietly abandoned and discontinued[259] due to high costs and technical challenges. Specifically, Loon's vision of stratospheric balloons delivering Internet

[253] https://www.forbes.com/councils/forbestechcouncil/2024/09/04/leveraging-low-earth-orbit-satellite-constellations-to-bridge-the-digital-divide/
[254] Nwana, H. S. (2018), HAPS for Affordable Broadband Connectivity, May 2018, from https://cenerva.com/reports/
[255] ibid
[256] HAPS aircrafts possess rechargeable power systems using solar power to fly during the day and high-capacity Lithium LI-ion batteries for use at night.
[257] https://en.wikipedia.org/wiki/Facebook_Aquila
[258] https://en.wikipedia.org/wiki/Loon_LLC
[259] https://www.bbc.co.uk/news/technology-44624702

faced technical, scalability and cost hurdles. For example, Loon technically works using "free-space optical communications" (FSOC), essentially fibre-optic Internet without the cables. A standalone company called Taara emerged from Alphabet's Project Loon in 2017[260] – spun out of Alphabet's X labs – focusing on bridging connectivity gaps worldwide. It uses innovative FSOC technology to transmit high-speed Internet wirelessly through beams of light. Taara has successfully deployed its technology in over a dozen countries, including Kenya, India, Ghana, and Fiji[261], and has introduced ground-breaking advancements like the Taara chip[262] that miniaturises its optical communication system. In Kenya, Poa Internet[263], an ISP, deployed Taara revealing unexpected bandwidth reselling by customers leading to the development of "Taara Share" which allows local entrepreneurs to become ISPs, offering affordable connectivity to their communities through pay-as-you-go (PAYG) transactions. Another notable development regarding HAPS is the involvement of Japanese corporation SoftBank. In June 2016, SoftBank Corp invested $15 million in the US-based start-up Sceye, which manufactures zeppelin-like airships[264].The Japanese company aims to launch pre-commercial mobile service in 2026 from High-Altitude Platform Stations (HAPS) operating in the stratosphere.

**HIBs** stands for "High Altitude Platform Stations as IMT Base Stations" (Euler *et al*, 2021). They are a subset of HAPS, since HIBs act as base stations for mobile networks, but ones which operate in the stratosphere [rather than terrestrially on the ground] to provide communication services. The promise of HIBs is to complement terrestrial networks, thereby extending coverage to remote and rural regions. They use the same frequency bands as ground-based mobile stations, making them compatible with existing devices that people use daily.

---

[260] https://www.datacenterdynamics.com/en/news/taara-exits-alphabets-x-labs-secures-external-funding/
[261] https://www.silicon.co.uk/networks/broadband/alphabet-spins-outs-taara-to-challenge-musks-starlink-605046
[262] https://www.datacenterdynamics.com/en/news/taara-exits-alphabets-x-labs-secures-external-funding/
[263] https://cioafrica.co/kenyan-low-cost-isp-poa-internet-secure-28-million-from-series-c-funding/
[264] https://www.lightreading.com/6g/softbank-fast-tracks-haps-timeline-partners-with-sceye-for-2026-launch

***Summary on HABs & HIBS:*** my overall take is this: though HAPs, HIBs and other FSOC-based solutions offer promising solutions for bridging the digital divide, they are likely to be even more *niche* than LEO/MEO/GEO satellite solutions, especially in remote and underserved areas, or would be used for disaster scenarios mainly. They still face high deployment and maintenance costs, along with potential regulatory and logistical challenges too. Innovative FSOC-based solutions like Taara experience predictable problems due to the sheer physics of wave propagation, notably the long distances from space and the susceptibility of air-transmitted data to weather conditions like rain, dust, fog, and heat, as well as the need for a clear line of sight. Just maintaining precise alignment of narrow light beams on swaying infrastructure due to winds is a clear engineering challenge. Nevertheless, the creative Taara team has developed AI-driven mirror systems and predictive algorithms to rapidly align the beams using AI-based adaptive beam control systems that dynamically adjust laser power based on real-time conditions like fog, rain, sun or dust. Bottom line – such space-based solutions would never be technically superior to their terrestrial based counterparts.

vi. *Public Wi-Fi Initiatives*: many public Wi-Fi projects aimed at providing free Internet access have faced sustainability challenges, limited coverage, and security concerns. The challenges of sustainability, scalability, and funding for broader implementations of Wi-Fi have always dogged such public Wi-Fi initiatives.

vii. *TV White Space (TVWS) technologies:* this technology (see Nwana, 2014, Chapter 6) uses the gaps between TV broadcast radio frequencies (also known as geographically interleaved spectrum) to provide Internet access. While TVWS has shown promise with limited pilot deployments in perhaps a dozen to two dozen countries, regulatory and technical challenges – in addition to the lack of affordable devices and equipment – have restricted any widespread rollouts.

***My overall summary on all of these 'sexy' innovative connectivity solutions:*** my summary is that – sadly – all these 'sexy' innovative technologies (5G, 6G, satellite solutions including LEOs, HAPs, HIBs, FSOC-based solutions, TVWS and even public Wi-Fi initiatives) have hardly made any dent to the 2.6 billion not online.

There is much hype about LEO services in particular promising to address connectivity gaps in remote areas with their Direct-to-Device (D2D) or Direct-to-Cell (D2C) capabilities. However, through basic physics, they will always struggle to replicate terrestrial 2G/3G/4G/5G networks' capacity, reliability, and lower latencies. Therefore, though D2C/D2D solutions are emerging, they will highly likely be sub-scale, complementary, solutions rather that replacing terrestrial mobile networks.

Overall, these 'sexy' promising new solutions may help at the edges to address the 2 billion who are *not* meaningfully connected, as they are all subscale to current terrestrial networks. They will make *de minimis* contributions to connecting the tail 2.6 billion unconnected. This reality elicits the importance of finding realistic, sustainable and wider scale solutions, or cheaper solutions that can be replicated widely in numerous locations.

## 6.4   When Market Solutions Fall Short – What Now?

With market-led solutions failing to connect the remaining 2.6 billion, what is the next step? Let me summarise the logic of this chapter thus far to be able to proceed to my proposals on some possible next steps:

i.   Section 6.1 warns – in much detail – of the folly of waiting for solutions to the 2.6 billion *not* online today coming from the telecoms industry as we know it today. It does this by pointing out that telecoms market has failed key [rural] populations – even for basic 2G voice services. It points out how 175 million people in 2025 would have to travel to a place where there is even basic 2G coverage (the coverage gap problem) and somewhere, conservatively, in the range 1,294 – 2,000 million people could not effect a basic 2G call *regularly* (the usage gap problem). Given the latter, it is easy to see how the telecoms market will fail most of the tail 2.6 Billion not online even more. Section 6.1 concludes that the estimates of Unconnected and the not meaningfully-connected Internet Usage Gap is in the range of 4,355 - 5,030 million people, with a mid-point of 4,625 million people. This is at the heart of the title of this book: *The Connectivity Crisis - Half the World Left Behind.*

ii.    Section 6.2 makes the case that, because of the enormity of the challenge spelt out in Section 6.1, regulatory [and policy] *intervention* is therefore very necessary. As illustrated in Figure 6.4, when the market and competition between its participants fail to produce positive outcomes for consumers and citizens – a significant market failure – regulation becomes necessary to ensure net benefits. This regulation is depicted on the left side of Figure 6.4.

iii.   Section 6.3 warns the audiences of this book to beware of the hype and promises of [new] sexy technology solutions, by reviewing them to reveal that all of them will only provide *niche solutions* to the significant market failures that I point out in Section 6.2.

Given these three core conclusions, the following questions and/or next steps emerge:

1.    *What* kind of technical [intervention] solutions are required? Chapters 4 and 5 provide my views on this question.

2.    *What* kind of regulatory and policy [intervention] solutions are required? This is the subject of Chapter 8 in Part III (of this book) that follows the next chapter.

3.    Given the enormity of the task as covered in Sections 6.1 and 6.2, *Who* should take a lead on these technical and regulatory interventions – and *How* do we ensure the chances of success are maximised? I think true leadership is an absolute must, and this is the subject of Chapter 9.

I believe these questions are critical ones – which would help map out some key approaches to address such enormous voice and Internet market failures affecting billions, mostly in LDCs, LICs and LMICs.

## 6.5   Summary

There are two core warnings that I convey in this chapter. First, that the telecoms market as we know it today in mid-2025 has already failed some number of humanity between 1.449 billion to 2.175 billion people globally, who could not *regularly* make a basic 2G voice call. Second, that the same telecoms market will largely fail to connect the tail 2.6 billion unconnected to the Internet. I also point out in this chapter that circa 2 billion of those that are connected today, are hardly meaningfully connected. This cumulatively makes more than half of humanity unconnected or under-connected, hence the title of this book *The Connectivity Crisis - Half the World Left Behind*. I also warn against the hype of several 'sexy' technology solutions that will never address these market failures across both basic 2G voice services as well as Internet connectivity. I conclude that regulatory intervention is very necessary going forward.

# Part III – Strategy, Collaboration, Policy, Regulation & Leadership

This **Part III** of the book consists of four chapters.

The first (Chapter 7) is on *Strategy & Collaboration*: attempting to address the broad issues of *individual country* strategies as well the *inter-country* collaboration strategies – that would drive up the connectivity amongst the 2.6 billion of people *not* connected to the Internet as of early 2025, and help meaningfully connect 2 billion others.

The second (Chapter 8) is on *Policy & Regulation*: attempting to address clear policy and regulatory recommendations that would/may mitigate the connectivity digital divide amongst the 2.6 billion of people *not* connected to the Internet.

The third (Chapter 9) is on *Leadership*: this is the chapter that gives this book its sub-title: "Expert Insights for Global Institutions and Developing Nations Committed to Achieving Universal Digital Inclusion". It overviews many categories of leadership that are all involved in some way in bridging the connectivity digital divide across the globe, not least because it gets quite confusing.

The last chapter briefly summarises the entire book.

So, this final part of the book addresses some enormously important aspects to realising the online digital inclusion challenge as well as the achieving more meaningful connectivity.

# 7  Strategy & Collaboration: Closing the Digital Divide

The long form title of this chapter is this – Strategy and Collaboration towards mitigating the digital divide amongst poorer nations. This chapter attempts to answer the following broad questions: what sort of *individual country* strategies and *inter-country* collaborations would truly make a dent to the massive 2.6 billion of people *not* connected to the Internet, and help meaningfully connect 2 billion others? These are self-evidently *monumental tasks*.

However, there is the much bigger problem. There are too many proverbial 'cooks in the kitchen', but each of the 'cooks[265]' are far too subscale to the size of the task of connecting the tail 2.6 billion and meaningfully connecting 2 billion others. This is due to something regulators and policymakers call *coordination market failure*. In this specific context, coordination market failures occur because it is too costly for individual national regulators, individual national universal service funds (USFs), individual or even collective major national telecom operators, and even LMIC/LIC/LDC countries themselves to *individually* and efficiently address their own Internet connectivity market failures.

For example, it is implausible for them individually to efficiently negotiate with global mobile technology firms like Huawei, Ericsson, or Nokia – for the latter to supply them with the cost-effective technology solutions they need *for rural needs*. They are too fragmented, numbering into hundreds – if not thousands – of LMIC/LIC/LDC countries, regulators, ICT Ministries, USFs, other well-meaning digital divide stakeholder players (cf. Section 1.5 in Chapter 1), etc. – collectively. They are individually too subscale too to address their connectivity needs effectively.

In addition to the coordination market failure are the even more significant demand-side market failures too, discussed in Parts I and II of this book. And

---

[265] Read 'individual LMIC, LIC or LDC countries', 'individual Aid agencies' – cf. Section 1.5, 'individual technology providers', 'individual Universal Service Fund (USF)', even the ITU itself, etc.

these demand-side challenges like [crushing] poverty, illiteracy, poor digital literacy and more tend to be national – are much more intractable to address. Add to these, the significant supply-side problems too.

Therefore, it appears to me that there is a logical and dire need for two core approaches and broad top-level strategies:

1. **Country-led strategy**: first, a fit-for-purpose and *individual* country-led strategy to optimise its digital divide for both voice and data, whilst addressing the demand and supply-side challenges.
2. **Collaboration strategies**: second, to address the coordination market failure, the individual LMIC/LIC/LDC/EDC countries must collaborate amongst themselves and with other stakeholders. There is a clear need for strategies or a main framework for these hundreds of fragmented digital divide protagonists *to coordinate and collaborate their sub-scale efforts.*

The rest of this chapter opines on these two strategy and collaboration challenges, i.e. *intra-country* strategy and *inter-country* collaboration strategies. This chapter outlines the lessons – and learnings – from the successful EDC India BharatNet programme, highlighting both positive and negative aspects of the Indian rural connectivity programme.

## 7.1 Crafting a Country-led Mitigating Digital Divide Strategy

There is nothing in what I articulate in this section that is truly novel. With only 35% of the population in LDCs using the Internet as of early 2025, particularly in Africa, clear country-led strategies and targets are needed to address this issue. It is evident that action is required. Lower-middle-income countries (LMICs), low-income countries (LICs), and least developed countries (LDCs) must address the digital divide with greater seriousness, rather than merely acknowledging these challenges without substantial efforts. So, these country-led strategies, plans and targets must be 'living' ones too.

Rather in many of these countries, so called 'broadband plans' are developed as box-ticking exercises. There needs to be a serious over-arching *Mitigating Digital Divide Strategy* for each LDC, LMIC or LIC which is truly idiosyncratic to the specific country. Clearly, the mitigating digital divide strategy for PNG (cf. Section 2.2) will be different – and should be different – from the equivalent strategy for Malawi (cf. Section 2.3). This should be obvious. However, I seldom observe well-constructed 'living' strategies for mitigating the digital divide in any of these countries. Furthermore, there appears to be a lack of comprehensive, actionable plans that support digital divide mitigation efforts and targets in most Least Developed Countries (LDCs), Low and Middle-Income Countries (LMICs), or Low-Income Countries (LICs).

### 7.1.1 Elements of a Mitigating Digital Divide Strategy

A comprehensive mitigating digital divide strategy should squarely address both demand-side and supply-side challenges to ensure equitable access to digital technologies and the Internet. The strategy must also be 'living' and internally coordinated, including monitoring and evaluation targets.

On the *supply side*, the 'living' strategy should address the following:

- *Policy Regulatory Frameworks:* each LIC/LMIC/LDC/EDC country must set up a clear and supportive policy and regulatory environment (cf. Chapter 8) that would encourage investment and innovation in both their voice and data digital sub-sectors.
- *Digital Infrastructure Development including addressing power challenges*: the strategy needs to address how to promote investments (both by the public/Government and private sector players) in building and upgrading digital infrastructures, like fibre backbones, 2G voice networks, 3G/4G/FWA broadband networks, mobile towers, satellite networks, and even internet exchange points (IXPs). The infrastructure development needs to be based on a living electrification strategy too.
- *Driving Digital Demand using Digital Capacity Building*: I personally find it very frustrating to see LICs/LDCs/EDCs professing 'digital', yet they hardly promote and drive the digitalisation and digitisation[266] of Government services. Digital technologies are revolutionizing Government

---

[266] Digitisation focuses on converting and recording data digitally (including today's paper-only records), whilst digitalisation concerns developing processes and changing workflows to improve manual systems.

services in some leading LDCs with good to excellent ICT leadership. For example, Rwanda [which suffered a terrible genocide 30 years ago] is emerging as a leader among Least Developed Countries (LDCs) in digital and e-Government. On or before 2022, Rwanda had already offered 98 e-Government online services[267], reflecting Rwanda's robust commitment to leveraging digital tools to improve governance and public service delivery. Education is another obvious sector that LDC Governments could use to drive digital demand, since digital divide in education is particularly unfair and egregious as it greatly disadvantages rural schools (primary and secondary) with no Internet connectivity, compared to urban schools with them. The provision of digital training and resources within the Government sector and for local businesses would help drive the necessary demand for digital voice and data services in LDCs.

- *Affordability*: affordability is both a supply and demand side challenge. On the supply side, Governments and regulators should implement policies and regulations to reduce the cost of voice and data/Internet access, including USF interventions, subsidies, tax incentives, and public-private partnerships.

On the *demand side*, the 'living' mitigating digital divide strategy should address, as best as it can, the following:

- *The perennial problems of poverty and lack of education:* these perennial problems can truly and only be addressed by Government. The uneducated and poor population are highly unlikely to be consumers of basic voice services, yet alone data/Internet services – particularly if digital devices like feature phones and smartphones remain completely unaffordable to the poor. Addressing the voice and data/Internet digital divide provide additional excellent reasons and motivations for reinforcing efforts to address poverty and educating all/most of the country's citizenry.
- *Digital literacy*: general literacy is one thing – digital literacy is another. It is important for any such mitigating digital divide strategy to encourage the development of programs to educate and train SMEs and citizens in digital skills, encouraging and assuring the effective use of digital technologies.

---

[267] https://desapublications.un.org/sites/default/files/publications/2022-09/Report%20without%20annexes.pdf

- *Digital Awareness Campaigns & Inclusion:* this is the demand-side's equivalent of the supply side's 'Driving Digital Demand using Digital Capacity Building'. The strategy must include the development of digital awareness campaigns to highlight the benefits of digital inclusion and encourage adoption among all demographics in society and SMEs. The strategy must also ensure inclusion including rural dwellers, the poor, the marginalised, underserved groups/areas, women and youth.
- *Services & localised relevant content:* the strategy must promote the creation and availability of locally relevant digital services and relevant local content that meet the needs of the LDC consumers and citizens.

The mitigating digital divide strategy must be truly *living* and *internally coordinated and coherent*:

- *'Living' and internally coordinated strategy:* what truly is the point of having a mitigating digital divide strategy which is not a 'living' one? A strategy that guides digital plans and activities daily, weekly, monthly and annually? The voice digital divide sub-strategy must be internally consistent and coordinated with a data/Internet sub-strategy, e.g. building out data/Internet infrastructure should effectively ensure it provides for voice services too.
- *Policy & Regulatory Coordination:* the lack of coordination I see in most LDCs between digital policy making [led by ICT sector Ministers] and digital regulation [led by 'independent' NRAs] is massive. So, the strategy should promote and ensure that digital divide strategies are integrated into broader national Government development policies.
- *Government and Regulatory Agency Coordination & Support:* For example, Governments and regulators can play a crucial role in facilitating seamless roaming agreements between Mobile Network Operators (MNOs), whilst ensuring fair competition and consumer protection.
- *Industry Collaboration:* using the example of roaming once more, telecom operators must collaborate and establish agreements to enable seamless roaming. The latter can be facilitated by industry standards and regulatory frameworks.
- *Monitoring and Evaluation:* the strategy must propose robust SMART Monitoring and Evaluation (M&E) targets and the necessary mechanisms to track and evaluate progress, as well as identify areas for improvement.

- *Public-Private Partnerships*: the strategy should encourage and foster collaboration between the private sector, NGOs and the public sector/Governments, to optimise and leverage all available digital resources and expertise.

Yet another rhetorical question – how many LDCs, LMICs or LICs possess *any* living mitigating digital divide strategy, yet alone a comprehensive one with all the above supply side, demand side and integrated features? Without such a true living strategy, the country would hardly make a dent into its voice and data digital divide challenges.

### 7.1.2  Evolving Broadband Plans to Bridge the Digital Divide

Broadband plans should evolve into much broader mitigating digital divide plans. Most LDCs and/or LICs/LMICs – under 'pressure' or 'obligation' from organisations like the World Bank or the ITU Broadband Commission – have developed – and some are implementing – 'broadband plans'. This is because without these plans, these countries would *not* be able to secure funding from these ilk of multi-national organisations like the World Bank and International Finance Corporation (IFC).

Broadband plans are generally designed to improve supply-side broadband digital infrastructure, enhance broadband connectivity, and foster economic growth in countries. Such broadband plan initiatives also enable LDCs/LICs/LMICs to access – not only financial support – but technical assistance, regional and global partnerships that would otherwise be unavailable.

In my experience, LICs, LMICs, and LDCs often treat broadband plans as formalities to satisfy the ITU and international funders, rather than genuinely implementing them. I would not name countries here including the one I hail from, but most of these broadband plans in LDCs/LICs/LMICs are hardly guiding broadband/Internet supply-side infrastructure development and/or demand-side adoption efforts. This means they are hardly 'living' and 'breathing' strategy documents. This is a true pity, because these LDCs/LICs are fooling themselves only, and the joke is on them and their citizens.

Not only should they take these broadband plans seriously, but they should enhance and evolve them into 'living' and 'breathing' 'Mitigating Digital Divide

Plans'. This is because – as the last chapter explains – we still have a huge usage gap of circa 1.3 billion minimum *not* using basic 2G voice services (cf. Section 6.1.1). So, the issue is not just broadband with LDCs/LICs/LMICs, but basic voice too. The broader Mitigating Digital Divide Strategy/plans should encompass today's broadband plans, as well as mitigating basic voice digital divide challenges.

These broader plans should not only be carefully crafted, but should be effectively implemented and tailored to local needs. They should be bolstered by stronger policy frameworks and capacity-building efforts to fully leverage the benefits of digital. Issues such as weak digital literacy, unaffordability of devices, and lack of relevant local content can hinder the productive use of broadband.

## 7.2 Lessons from India's BharatNet for Rural Connectivity Strategy

I thought it important to review a 'best practice' intra-Country digital divide mitigating strategy and implementation plan for widescale rural connectivity – India's BharatNet programme. It is important to present such a 'best-in-class' example of a country's rural connectivity [or bridging the digital divide] strategy/case study to illustrate the key components of an effective 'mitigating digital divide strategy' as detailed in Section 7.1.1. It is also important that LDCs/LMICs/LICs realise that they can truly bridge the digital divide if they put their minds to it. India has been profoundly serious about bridging its digital divide between urban and rural areas for more than a decade now.

I have quietly been extremely impressed by India's efforts and outcomes over the past fifteen years in addressing its rural connectivity challenges. Despite the shortcomings of the India BharatNet rural connectivity programme to date, its impressive outcomes are unarguable. The rest of this Section 7.2 draws from some joint work I did with Dr Shruthi Koratagere Anantha Kumar[268].

India's BharatNet programme represents one of the world's largest rural connectivity initiatives, aiming to bridge the digital divide across the nation's

---

[268] https://scholar.google.com/citations?user=JQdckh0AAAAJ&hl=en

vast territory. Since 2010, the programme has contributed to a remarkable increase to 900 million Internet users by 2025 across India. By all accounts, this is an incredible success that India has realised, a remarkable achievement for the Department of Technology (DoT), leading to a significant digital transformation across India. Looking at the sheer outcomes to date, India must be applauded – and I strongly recommend that dozens of other LICs, LLDCs and LMICs should seek to emulate and learn from India.

However, even India's BharatNet programme is finding out that Last Mile Connectivity and Usage Gaps are the key limiting factors to its gargantuan Internet connectivity efforts. India – as an emerging and developing country (EDC) – also has the capability to lead, collaborate and share its experiences with other LICs, LDCs, LMICs and EDCs to facilitate their equivalent rural connectivity initiatives.

The rest of this entire Section 7.2 which describes India's BharatNet programme is very largely a slightly revised version of a joint opinion piece previously published by Dr Shruthi Koratagere Anantha Kumar[269] (as principal author of the opinion piece) and the author of this book (as coauthor), published by Cenerva Ltd on its website[270]. It is largely republished here with the full permission of Dr Kumar, and clearly and suitably highlighted to clear reflect 'joint work' with her on this BharatNet 'Case Study' reported in the rest of this section. I have only made some minor changes and modifications on our previously published opinion editorial.

### 7.2.1   Introducing the BharatNet Connectivity Programme

India's BharatNet connectivity programme is arguably the world's largest rural connectivity initiative. The programme was first set up in 2010 to bridge the digital divide across India. In the early 2010s, digital adoption in India was low, with total Internet users around 92 million[271]. Launched in the early 2010s when Internet penetration was merely 92 million users, the programme has contributed to a remarkable increase to 900 million users by 2025. How did

---

[269] *Ibid.*
[270] https://cenerva.com/wp-content/uploads/2025/02/BharatNet-report.pdf
[271] "Internet users by region and country, 2010-2016." Accessed: Feb. 10, 2025. [Online]. Available: https://www.itu.int/en/ITU-D/Statistics/Pages/stat/Treemap.aspx

India achieve this? How successful has the connectivity programme been to date?

To encourage digital adoption of services in remote and rural areas, the Government of India designed and launched the first phase, **Phase I**, of its flagship programme called *BharatNet*, funded by the Indian Universal Service Obligation Fund (USOF) and administered by the Department of Telecommunication (DoT)[272] [273]. The USOF had been created to promote equitable access to telecommunication services across the country, particularly to bridge the digital divide in remote and rural areas. The key vision of the USOF programme is/was:

- Provisioning of quality and affordable mobile and digital services
- Allowing non-discriminatory access to mobile and network services
- Providing equitable access to knowledge and information dissemination
- Promoting rapid socio-economic development with improved standard of living

The ambitious project aimed (still aims) to connect 2.5 lakh [i.e. 250,000] Gram Panchayats (GPs) or village councils, through a comprehensive broadband Internet network infrastructure which would enable the launch of several digital services such as e-health, e-education, e-governance, e-commerce and e-finance in rural areas. BharatNet is implemented and managed under Bharat Broadband Network Limited (BBNL), which is responsible for planning, deploying and managing the BharatNet project. BBNL needs to ensure effective execution of the objectives. This enables access to service providers such as Mobile Network Operators (MNO), Internet Service Providers (ISPs), cable TV operators and content providers.

In 2017, India launched **Phase II** of the BharatNet programme to digitally empower every citizen of India, which it called 'Digital India'. The programme

---

[272] "USOF Ongoing Schemes - Advancing Telecom Connectivity in India." Accessed: Feb. 10, 2025. [Online]. Available: https://usof.gov.in/en/ongoing-schemes
[273] "BharatNet: Bridging the Digital Divide." [Online]. Available: https://pib.gov.in/PressReleseDetailm.aspx?PRID=2086701&reg=3&lang=1

explores ways to bridge the gap between those connected to the Internet and those who were not. The key vision areas of this Phase of the programme were the following[274]:

- Digital infrastructure as a core utility for every citizen – a supply side challenge;
- Digital empowerment of citizens – a demand-side challenge;
- Governance and coordination of digital inclusion stakeholders; and
- Development of e-Services on demand – an Internet demand stimulation exercise towards Indians using services such as e-health, e-education, e-governance, e-commerce and e-finance in rural areas.

As part of the BharatNet Phase II programme, network utilisation and revenue streams are generated through leasing lit fibre bandwidth and dark fibre, providing Wi-Fi access broadband services, Internet services in public places and Fibre-To-The-Home (FTTH) services. BharatNet has since entered **Phase III**: BharatNet Phase III officially began on August 14, 2024[275].

### 7.2.2  BharatNet's Impressive Results to Date (to 2025)

As of early 2025, around 900 million users are using the Internet in India. This is a remarkable achievement for the DoT, leading to significant digital transformation in India compared to fifteen years ago. Key statistics and data for India's BharatNet programme as of September 2024 are noteworthy[276]:

- Length of Optical Fibre Cable (OFC) laid: 683,175 km
- Number of Gram Panchayats (GPs) where OFC is laid: 210,552
- Number of GPs connected via Satellite backhaul: 4,952
- Total number of villages where connectivity would be provided through extended BharatNet coverage: 655,968
- Current coverage of villages through extended BharatNet coverage: 199,655

[274] "Digital India." Accessed: Feb. 10, 2025. [Online]. Available: https://csc.gov.in/digitalIndia
[275] https://bharatnet.in/bharatnet-phase-iii-landmark-initiative-narrowing-the-digital-gap-between-urban-and-rural-india/
[276] "BBNL, Ministry of Communications & Information Technology, Govt. of India." Accessed: Feb. 10, 2025. [Online]. Available: https://bbnl.nic.in/

- Wi-Fi hotspot in GPs installed: 104,574; <u>however only 6,039 Wi-Fi hotspots are active</u>
- FTTH connections commissioned: 1,141,825
- Dark fibre (in km): 94,234
- Note: the wireline industry penetration in India is around 20%.

As these above statistics prove, the entire BharatNet programme – implemented over three phases since 2011 – has achieved significant milestones, including the laying of over 683,175 km of optical fibre cable and connecting more than 210,552 Gram Panchayats (GPs).

India has also come a long way with its 'first mile' submarine infrastructure. As of the end of 2023, India hosts around seventeen international subsea cables across seventeen landing stations, with a total lit capacity of these cables of circa 180 Tbps (terabits per second)[277].

### 7.2.3 BharatNet: *How* it was Implemented

Here are some key unique characteristics and steps that describe how the BharatNet programme was implemented:

- In 2011, BharatNet Phase-I commenced by providing broadband connectivity to GPs by linking block headquarters of GPs through the-then existing Optical Fibre Networks of Central Public Sector Undertakings (CPSUs) such as Bharat Sanchar Nigam Limited (BSNL), RailTel Corporation of India Limited (RailTel) and Power Grid Corporation of India Limited (PGCIL), by laying incremental fibre to bridge the connectivity gaps up to the GPs.
- The BharatNet programme employs multiple implementation models, including *State-led, public sector/CPSU, private sector, and satellite connectivity* approaches. For example:

---

[277] "New undersea cables activation in 2025 to push India's data transmission capacity four times: Trai Chairman - The Hindu." Accessed: Feb. 21, 2025. [Online]. Available: https://www.thehindu.com/sci-tech/technology/new-undersea-cables-activation-in-2025-to-push-indias-data-transmission-capacity-four-times-trai-chairman/article69110409.ece

- o a State-led model has been implemented in eight Indian states to date including Chhattisgarh, Gujarat and Andhra Pradesh.
- o A public/CPSU[278]-led model has been implemented by BSNL in four Indian states to date including Madhya Pradesh and Uttar Pradesh.
- o A private sector model has been adopted in Punjab and Bihar.
- o A satellite-led model executed by BSNL and BBNL has connected 5,161 GPs via satellite to date in early 2025.
- In 2022, the Indian Government executed a strategic merger between BBNL and BSNL in order to streamline operations and accelerate deployments further. *There is a key lesson here.* I see so many digital divide overlapping agencies in LDCs falling over each other. These should be addressed as India addressed its own overlapping agencies.
- BharatNet has also adopted a carefully composed mix of technologies (fibre, radio, satellite). Last-mile connectivity was assured via Wi-Fi or other broadband technologies, including FWA.
- USOF evolved into Digital Bharat Nidhi (DBN) in 2024, which signals the India Government's renewed commitment to digital inclusion.

Commencing in August 2024, the final Phase III of the BharatNet Project aims to deliver high-speed Internet access to all villages in India, thereby enhancing connectivity into underserved regions. The focus of this phase is to strengthen and expand the existing network infrastructure laid in Phase I and Phase II. This phase also explores last-mile connectivity challenges *through collaboration with local village entrepreneurs* working with BSNL.

### 7.2.4 BharatNet Challenges – Lessons for Other Countries

Whilst it is important to acknowledge the positive aspects of the BharatNet programme, it is also necessary to recognise the significant challenges that the programme (still) faces.

Despite the impressive achievements enumerated in Section 7.2.2, the programme has faced substantial challenges. There are several reasons which

---

[278] Central Public Sector Undertakings

led to slowdown in the implementation of the BharatNet project that other LDC, LIC, LMIC and EDC nations must learn from. These challenges include implementation delays, infrastructure difficulties in remote areas, coordination issues among multiple stakeholders, operational inefficiencies, and competition from emerging technologies.

- *Implementation*: BharatNet has faced multiple delays due to challenges in planning, coordination and execution, which led to projects overrunning the budget and delivering incomplete infrastructures.

- *Infrastructure and technology*: the terrain, lack of proper roads and logistical difficulties in deployment are some of the major challenges that resulted in fibre not reaching some intended GPs. As the project has run over many years, some equipment became outdated, resulting in poor performance[279]. The optical fibre is expected to provide a minimum of 100 Mbps per GP, which is inadequate for a wide range of digital services and for scenarios where each GP consists of a cluster of villages[280]. There are also technical challenges in terms of end-user signal quality, bandwidth limitations and network downtime[281].

- *Coordination gaps*: there are multiple digital divide stakeholders involved in the BharatNet project, including central and state Governments, equipment providers, infrastructure providers, MNOs, ISPs, local bodies, private contractors, regulators and policymakers. There is a need for a more collaborative approach to improve coordination and efficiency in order to expedite the bottlenecks in infrastructure deployment.

- *Operational inefficiencies*: while electricity access is available for approximately 99% of the population in India, there remain challenges in guaranteeing reliable, uninterrupted power supply with high quality across

---

[279] Many other LICs, LMICs and LDCs will recognise this too.
[280] "BBNL FAQ." [Online]. Available: https://bbnl.nic.in/FAQ.aspx?pid=52
[281] "BharatNet Project." - https://www.drishtiias.com/daily-updates/daily-news-analysis/bharatnet-project-2

all areas[282] [283]. Many rural and remote areas still suffer outages, frequent fluctuations and dependence on diesel generators due to poor infrastructure and distribution.

- *Budget constraints*: there are significant budget constraints, which are affecting the project's Quality of Service (QoS) and scalability. The high operational cost of maintaining the fibre in remote rural areas is making this OFC backhaul less sustainable.

- *Competition from alternative technologies*: emerging technologies such as 4G/5G broadband services, Low Earth Orbit (LEO) satellites and wireless broadband technologies can provide faster and more cost-effective solutions, which are competing with BharatNet directly[284].

- *Lack of local engagement, governance and corruption*: Another critical challenge faced by the BharatNet initiative is the lack of local skilled or trained workforce who could engage private players in providing last-mile access. Village Local Entrepreneurs (VLEs) would play a critical role in the scalability of the rollout. There is also a need to monitor fund allocation and usage in order to minimise corruption and improve the credibility of this project.

[282] "BBNL, Ministry of Communications & Information Technology, Govt. of India." Accessed: Feb. 10, 2025. [Online]. Available: https://bbnl.nic.in/
[283] "Access to electricity (% of population)" https://data.worldbank.org/indicator/EG.ELC.ACCS.ZS?locations=IN
[284] "BharatNet Project." Available: https://www.drishtiias.com/daily-updates/daily-news-analysis/bharatnet-project-2

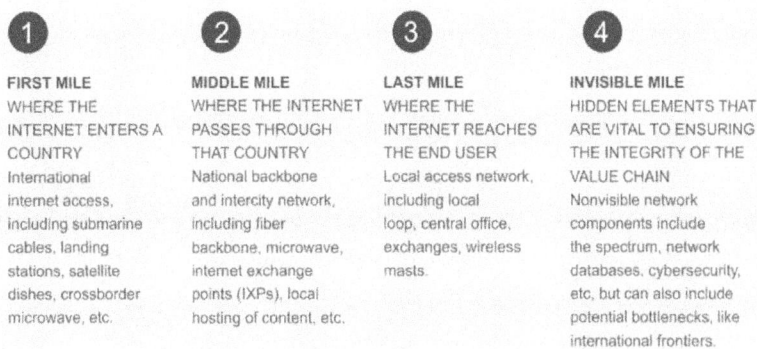

| ① FIRST MILE | ② MIDDLE MILE | ③ LAST MILE | ④ INVISIBLE MILE |
|---|---|---|---|
| WHERE THE INTERNET ENTERS A COUNTRY | WHERE THE INTERNET PASSES THROUGH THAT COUNTRY | WHERE THE INTERNET REACHES THE END USER | HIDDEN ELEMENTS THAT ARE VITAL TO ENSURING THE INTEGRITY OF THE VALUE CHAIN |
| International internet access, including submarine cables, landing stations, satellite dishes, crossborder microwave, etc. | National backbone and intercity network, including fiber backbone, microwave, internet exchange points (IXPs), local hosting of content, etc. | Local access network, including local loop, central office, exchanges, wireless masts. | Nonvisible network components include the spectrum, network databases, cybersecurity, etc, but can also include potential bottlenecks, like international frontiers. |

Figure 7.1 – Defining First, Middle, Last & Invisible Miles
Source (Adapted from World Bank[285])

The above challenges enumerated, we (Dr Shruthi K.A. Kumar and I) discerned and opined in our opinion piece[286] that the biggest challenges of the BharatNet Project today (in 2025) are *last-mile connectivity issues* (see Figure 7.1) and *low usage rates*.

1.  ***Last Mile Connectivity (Coverage Gap) Challenges***: the *last mile* is where the Internet reaches the end-user, in contrast to the 'first' and 'middle' miles (cf. Figure 7.1). India has come a long way with its the 'first mile'[287] infrastructure with its submarine connected capacity. As of the end of 2023, India hosts around seventeen international subsea cables across seventeen landing stations, with a total lit capacity of these cables at circa 180 TBps (terabits per second)[288]. India's 'middle mile' backbone fibre saw an increase in the inventory of operational terrestrial backbone optical fibre cable (OFC) laid, reaching 683,175 km by end September 2024, dozens times the figure of ten years ago. Once high-speed Internet reaches the centre of a GP village through the first and middle mile infrastructure,

---

[285] https://documents1.worldbank.org/curated/en/674601544534500678/pdf/Main-Report.pdf
[286] https://cenerva.com/wp-content/uploads/2025/02/BharatNet-report.pdf
[287] Terminology borrowed from the following report:
https://www.broadbandcommission.org/Documents/working-groups/DigitalMoonshotforAfrica_Report.pdf
[288] https://www.communicationstoday.co.in/indias-data-capacity-to-quadruple-with-new-submarine-cables-by-2025/

telecommunications operators deliver Internet services (including mobile and fixed internet services) to individuals, businesses, and Government entities through the last mile.

The last mile has presented untold challenges across the 250K GPs, and these last-mile connectivity challenges have contributed to the **coverage gap** problem. Recall, coverage gaps refer to populations that do *not* live within the footprint of a mobile broadband/Internet network. Along with steps like restoring electricity grids, integration of renewable energy and improving last-mile connectivity, the goal towards combining last mile connectivity and providing total electrification remains realisable in India.

2. **Low Utilisation (Usage Gap) Challenges** as shown in Section 7.2.2, the low utilisation challenge is evidenced by only 6,039 active Wi-Fi hotspots out of 104,574 installed in the rural connectivity BharatNet programme to September 2024. This is the **usage gap** problem: recall too that usage gaps refer to populations that live within the footprint of a mobile broadband/Internet network, but do who *not* use the Internet for a myriad of reasons covered at the beginning of Part II of this book. The demand side issues contribute to the usage gap challenge, wherein millions of people live within reach of a [mobile/FWA] broadband/Internet network but do *not* use the Internet.

So, in many GPs where there is last mile infrastructure, the lack of viable business models combined with low adoption of services has resulted in poor Returns on Investment (ROIs). This has led to predicted poor RoIs in other GPs, which in turn have mitigated further last mile infrastructure rollouts in these GPs. Sometimes, these are due to unreliable electricity in these regions/GPs, further mitigating last-mile connectivity solutions, digital literacy challenges, unaffordable pricing plans to subscribe for Internet services and others.

Nevertheless, I think India's BharatNet programme provides a true, scale and realistic mitigating digital divide strategy, and the positives and shortcomings of the programme provide clear lessons to other 'poor' nations.

### 7.2.5  Techno-Economic Modelling Would Help BharatNet

I am reasonably confident that the DoT would be able to address the former group of challenges of the previous section, including implementation delays, infrastructure difficulties in remote areas, coordination issues among multiple stakeholders, operational inefficiencies, budget constraints and competition from emerging technologies. It is the latter two, i.e. the last mile coverage gap and usage gap challenges, which are the most acute.

India has a population of at least 1.44 Billion people[289] as of early 2025, and realising the connectivity outcomes for the remaining *third* 500M people of using the Internet would be exponentially harder – recall that the Pareto principle (80/20) rule of Chapter 1 predicts this.

This calls for DoT to be much SMART-ER[290] /SMARTER in its efforts going forward. We propose that DoT employs more fine-grained, Geographical Information System (GIS)-based technical-economic modelling as an approach to inform all of its next steps. Techno-economic modelling for rural broadband would provide a more *transparent* and *comprehensive* approach that evaluates both the technical and economic aspects of deploying broadband Internet networks in all of the 2.5 lakh GP rural areas. This method helps in identifying the most cost-effective and efficient solutions for providing high-speed Internet access to these GP regions.

The core tenets of techno-economics include:

- *Technical Assessment:* evaluate, in particular, last mile broadband technologies (e.g., fibre optics, satellite, wireless, FWA, TVWS, etc) for feasibility and performance in rural areas, considering coverage, bandwidth, and reliability.

- *Economic Analyses*: this part evaluates the expenses involved in deploying and maintaining broadband Internet infrastructure. It includes capital

---

[289] https://data.worldbank.org/indicator/SP.POP.TOTL?locations=IN
[290] Specific, Measurable, Actionable, Realistic, Time-Bound, Ethical and Reviewable

expenditures (CapEx) for equipment and installation, as well as operational expenditures (OpEx) for ongoing maintenance and service provision.

- *Regulatory and Policy Considerations*: it is essential to comprehend the regulatory environment and Government policies. This entails examining subsidies, grants, and other financial incentives that can facilitate rural broadband projects. The policies should be technology-neutral and should also be flexible enough to accommodate new ways of providing backhaul (for example, low Earth orbit satellites), or last mile solutions (for example, TV white spaces). This can include policies and regulatory frameworks that are subject to periodic review.

- *Market Demand and Adoption*: it is essential to estimate the potential user base and their willingness to pay for broadband/Internet services within each GP. This requires conducting thorough market research to understand the needs and preferences of rural communities.

- *Cost-Benefit Analysis*: comparing costs and benefits of broadband solutions helps stakeholders choose the best options and make informed investment decisions.

- *Case Studies and Scenarios*: real-world examples and hypothetical scenarios test model feasibility, offering insights into the challenges and successes of rural broadband projects.

We recommend that the BharatNet programme should build all of the above into a transparent Rural Connectivity Portal based on the detailed approach depicted in Figure 8.3 (in the next chapter). Such a detailed approach systematically and accurately identifies coverage gaps, generates population coverage statistics, estimates unmet demand, forecasts future demand and calculates CAPEX and OPEX requirements for all infrastructures/sites rollout. The Rural Connectivity Portal should be capable of being expanded to include low cost solutions for Internet access and different types of technologies. It could also be used to inform the right interventions to address demand-side challenges such as lack of awareness, lack of digital skills, exorbitant usage-pricing models and more.

The reader is encouraged to read Section 8.3.4 of the book in conjunction with this section.

## 7.3  National and International Collaborations for Digital Inclusion

What I am really referring to in this section are both the *intra-country* and *inter-countries* collaboration strategies required for mitigating the digital divide. Recall that at the start of this chapter, I note that there is a massive coordination failure at play amongst the multiple dozens of individual LMICs/LICs/LDCs, along with hundreds of other digital divide stakeholders, for them to efficiently address their digital divide challenges. I note that this is due to something regulators and policy makers call *coordination market failures*. It is a sad truism that many such failures occur intra-country. For example, in the BharatNet case study overviewed in the prior section, a significant challenging issue has been coordinating multiple internal stakeholders in India. As seen, the Indian Government decided to merge some key entities to enhance coordination and efficiency. Specifically, that, in 2022, India executed a strategic merger between BBNL and BSNL in order to streamline operations and accelerate deployments further.

So, it is important for LMIC/LIC/LDC/EDC nations to find a way to collaborate *both* among themselves and with other stakeholders, such as through intra-country collaborations. Therefore, there is a need for frameworks, or a main framework, to enable the many fragmented *international* digital divide participants, including inter-countries efforts, to coordinate and collaborate their efforts effectively and efficiently. In this vein, I propose the following.

### 7.3.1  Prioritise Intra-Country Mitigating Digital Divide Strategy

For completeness, it is worth noting that any inter-countries collaboration should proceed in tandem with a 'living' intra-country mitigating digital divide strategy, plan and roadmap. The plan should – at a minimum – abide by the elements or contents proposed in Section 7.1. Other nations should learn from India's BharatNet programme as covered in Section 7.2.

### 7.3.2 Why So Little Collaboration Happens to Date Amongst LDCs

Why LDC, LIC, LMIC, SIDS/Caribbean[291] and EDC nations do not collaborate fully with each other on addressing their respective digital divide challenges has always astounded me. They each feel the pain of their massive digital divide challenges, which the developed nations/regions like Australia, the EU, the USA, Canada, UK, Japan, Singapore, etc. do *not* feel. So why don't they collaborate? This question needs to be answered properly in order to *inform* ways of them collaborating.

I believe there are many reasons why LDCs, LICs, LMICs and even EDCs [like India] do *not* tend collaborate fully on addressing their digital divide challenges:

1. *Believing in, Waiting and Hoping for the 'Trickle-Down' Effect to Work – and Believing in Current Equipment Manufacturers*: many of these countries implicitly believe in their digital inclusion/divide solutions eventually coming from 'trickled-down' from developed nations. Or waiting for solutions from the Broadband Commission, the World Bank, the ITU and/or a myriad of others. And they also implicitly believe that current equipment manufacturers like Huawei and Starlink would come to their rescue – eventually. For example, Huawei (from China) is constructing 70% of Africa's 4G networks. Startups employing more affordable technologies are investigating ways to connect remote communities, and the likes of Starlink promise to connect rural areas too. *I hope what the reader has read so far to this point of this book should disavow them of these terribly misguided notions.* 'Trickle-down' hoping is not the approach to addressing the tail 2.6 billion unconnected.

2. *Lazy ICT Ministers, Policy Makers and Regulators*: this must be said out loud. I have met so many LIC/LMIC/LIC/EDC ICT ministers, policy makers and regulators over the last decade and a half who know better – that trickle-down and/or waiting/hoping for solutions to their digital divide woes from others but themselves is *not* working – and will *not* work. Yet, they lazily do nothing. This must be strongly called out.

---

[291] The United Nations categorises the Caribbean as Small Island Developing States (SIDS), a distinct group of nations which face unique social, economic and environmental vulnerabilities.

3. *Low Prioritisation of Digital Divide Issues and Resource Constraints*: many of these countries do not truly prioritise the digital divide as they have other more urgent pressing concerns. To be fair to them too, they also face significant financial and technical resource constraints. So, they prioritise immediate needs over long-term digital infrastructure and digital capacity building investments.

4. *Geopolitical & Colonial Factors*: Geopolitical considerations and colonial history can (and does) influence collaboration between countries. Cooperation of any type between neighbouring African countries like Ghana and Ivory Coast, Nigeria and Cameroon or South Sudan with Central African Republic have largely been non-existent because the latter countries of these pairs [i.e., Ivory Coast, Cameroon and Central African Republic] are all former French colonies – who are still in 2025 under the yoke of Paris. France still largely determines their international cooperation agendas – I have observed this first hand in my own country of origin, Cameroon. It is both disgusting and unbearable, but this is the reality we Africans living in *France Afrique* still bear with in 2025, 65 years after our so-called independences from France.

5. *Political and Economic Instabilities*: yes, political instability and economic challenges in many of these LDC countries hinder collaboration, e.g. DRC, South Sudan, Myanmar, etc. For example, my own native land Cameroon, as of 2025, is embroiled with the Boko Haram crisis[292] in the North of the country simultaneously with an Anglophone-Francophone civil war in the West of the country. Such instabilities put digital divide priorities on the back burner.

6. *Lack of other Basic Infrastructures*: the lack of other basic [utility] infrastructures, such as clean water grids, sewage systems and roads that can support the rollout of fibre backbones, electricity and Internet connectivity, makes it difficult to implement digital solutions. This is a common issue in many LDCs.

---

[292] Boko Haram is a jihadist militant group based in northeastern Nigeria, also active in Chad, Niger, northern Cameroon, and Mali.

7. *Diverse Country Needs and Priorities*: each country faces unique challenges and priorities, making it hard to find common ground for collaboration. So, Governments focus on internal navel-gazing on digital divide issues rather than on international cooperation.

8. *Etc.*

So, what tends to happen is that external stakeholders tend to attempt to broker digital divide collaborations amongst such LDC countries, e.g. the United Nations Development Programme (UNDP) [293], the GSMA[294] or the United Nations Technology Bank for the LDCs[295] who are working to strengthen digital capacities in these regions.

These LDC and EDC countries must learn to <u>collaborate directly</u> with each other, and these above 'reasons' for not collaborating must not be used as excuses. Good leadership is the only antidote I know to rise above these excuses.

### 7.3.3 Possible Inter-Countries Collaboration Strategies & Benefits

Once each LDC or SID has established their *intra-country* digital divide mitigating strategy, they should seek possible collaborations with other fellow LDCs and/or SIDs. Collaborations always help share the load (and the pains), and they could help bring results faster.

Observe how the leadership and collaborations amongst the 27 European Union (EU) nations in the domains of telecoms policy, regulation and investments over the past 30 years have undoubtedly benefitted *all* the countries with the EU. The European Union's leadership and collaborative approach driven by the European Commission (EC) has definitely played a significant role yielding benefits such as harmonised regulations and joint investments in digital infrastructure projects

---

[293] https://www.undp.org/blog/committing-bridging-digital-divide-least-developed-countries
[294] https://www.gsma.com/about-us/regions/sub-saharan-africa/wp-content/uploads/2023/10/USF-Africa.pdf
[295] https://www.un.org/technologybank/news/ldc-insight-4-strengthening-digital-capacities-least-developed-countries-even-more-urgent-post

such as the Connecting Europe Facility[296], have led to improved connectivity and digital services across the region. EU collaborations offer numerous advantages, including standardisation, consumer benefits like reduced roaming charges and enhanced consumer protection. The EU Digital Single Market has facilitated joint research and innovation through initiatives such as Horizon Europe[297]. Horizon Europe has contributed to the development of innovative technologies and digital solutions. Other regions might consider learning from the EU's experience and adapting similar strategies to their specific contexts.

Therefore, there are possible strategies to help LDCs, LICs, LMICs, SIDS and EDCs start collaborating on addressing their digital divide challenges – and some potential benefits:

1. *Set up regional digital divide partnerships*: the successes of the EU model above demonstrate the potential benefits of regional cooperation and collaboration in addressing digital divide challenges. Countries within the same region (e.g. sub-Saharan Africa, Caribbean, Pacific Islands, etc.) often share similar challenges and feel similar 'digital divide' pains. Forming regional alliances or forums can effectively consolidate resources and expertise. Organisations such as the African Union or ASEAN could play a pivotal role in facilitating these partnerships.

2. *Create Joint Initiatives*: collaborative projects between LDC/SIDS-type nations on bridging their digital divides on digital infrastructures, digital technologies, digital literacy programs, and technology transfer can help many LDC, LIC, LMIC and SIDS nations. These initiatives can get support from international organisations, NGOs, and private sector partners. Imagine a joint initiative led by EDC India to help some SIDS nations in the Asia Pacific region, such as the Pacific Island States.

3. *Share Digital Divide Best Practices*: countries can benefit from each other's experiences. Establishing platforms for knowledge exchange and sharing case studies can help spread effective strategies and solutions. Again, I

---

[296] https://commission.europa.eu/funding-tenders/find-funding/eu-funding-programmes/connecting-europe-facility_en
[297] https://commission.europa.eu/funding-tenders/find-funding/eu-funding-programmes/horizon-europe_en

know many countries in sub-Saharan Africa who would truly benefit from learning from India's BharatNet programme.

4. *Jointly Access and Leverage International Financing:* securing international financing from entities such as the World Bank, International Monetary Fund (IMF), and the United Nations may be easier to enable essential financial assistance for joint digital divide initiatives, rather than just for one country securing the finance. Accessing and leveraging international financing for joint initiatives between LDCs can be easier for several reasons: *economies of scale* by pooling resources to be more credible to attract the financing in the first place; *similar policy alignment*, e.g. to realise similar SDG ICT goals; *increased visibility and profile* which would demonstrate more seriousness to international financiers and NGOs; *shared and similar digital divide pains/challenges* such as poverty, low literacy, etc., leading to international donors seeing the value in collaborative efforts between two or more countries; *more efficient multilateral support* leading to organisations like the ITU or the UN providing technical assistance and capacity-building more efficiently, and more.

5. *Engage Private Sector Tech companies more efficiently:* consider the economies of scale effects of several LDC/EDC nations – through a joint initiative – partnering with technology companies such as Huawei, Ericsson or Starlink for better value commercial deals on equipment. Ditto with other private tech companies and investors who can bring in invaluable expertise, technology, and funding, e.g. Meta/Facebook, Google, Amazon or Microsoft. Such Public-Private Partnerships (PPPs) are clearly more efficient with a group of countries serious about addressing their digital divide challenges.

6. *Develop Standardised Policies and Regulations:* harmonised policies and regulations have been key to the success of the EU project across many spheres. Aligning policies and regulations across different regions and/or nations can enable smoother collaboration and execution of digital divide projects. This encompasses topics such as infrastructures procured, data privacy, cybersecurity and telecommunication standards.

7. *Foster Communities' Involvement:* this will drive up demand-side adoptions. Consider the joint LDCs initiatives ensuring and encouraging

their local SMEs[298], corporates and NGOs are engaged in the digital divide initiatives, including the engagements of local communities and stakeholders in the planning and execution of digital projects. Such engagements can also drive economies of scale and scope, ensuring that digital divide solutions are tailored to the specific needs of region and/or countries.

8. *Utilise Technology for Coordination*: digital tools and platforms can enhance communication and coordination among collaborating nations. Virtual meetings, online collaboration tools, and shared databases can significantly improve the efficiency of collaboration efforts.

Collaborations can and would truly yield more effective and sustainable solutions to address the challenges of the digital divide encountered by these countries. This necessitates a coordinated effort from Governments, international organisations, the private sector, and local communities.

## 7.4 India: A _Potential_ Champion for Nations Battling the Digital Divide Pains

India could lead a joint initiative of digital divide nations – she understands their pains. Put another way, India can lead the charge as she understands digital inequality firsthand and has been firmly addressing it through the BharatNet programme (cf. Section 7.2).

### 7.4.1 Why India? Leading Low-Cost Mobile Innovation

Why do I pick on India? India's push for digital inclusion is clear for all to see in this chapter. India can drive cheaper and more affordable mobile technology standards. India feels the digital divide pains of other LDCs, LICs, LMICs and other EDCs – which the USA, EU, Australia, Canada, the United Kingdom, etc., do *not* feel. Afterall, though India has incredibly connected 900 million to the Internet over the past fifteen years, another 500 million *are still to be* connected.

---

[298] Small and Medium Enterprises

In addition, I think India has shown that, if it puts its mind to addressing some major goals, it can realise them. Consider the following:

1. *The BharatNet programme*: despite its shortcomings, the programme to date demonstrates unarguable digital divide successes that speak for themselves.

2. *India's successful Chandrayaan-3 mission*: India has indeed made significant strides in space exploration, with its successful Chandrayaan-3 mission. On 23rd August 2023, India became only the 4th country to reach the moon by successfully landing its Vikram lander spacecraft near lunar south pole[299] [300]. "India is now on the Moon," announced Prime Minister Narendra Modi immediately after the Vikram lander touched down on the lunar surface. India has a rapidly growing economy, and technological advancements. India is a true emerging power with the potential to influence global economic dynamics.

   India's Chandrayaan-3 mission succeeded due to meticulous planning, resilience, and commitment. The Indian Space Research Organisation (ISRO) showed perseverance and dedication in space endeavours. Failures like Chandrayaan-2 were seen as learning experiences. ISRO analysed these failures, made improvements, and pushed forward with determination. So, India is capable of taking on setting itself hard goals and realising them. India can therefore take a significant *global* lead on bridging the digital divide – if she chooses to. The goodwill she will earn will be humongous.

3. *India has a robust 3G/4G manufacturing industry*: India is also preparing for the 5G era, with significant investments and initiatives aimed at making the country 5G-ready[301]. India is the world's second-largest telecommunications market. As of May 2024, the total telephone subscriber base stood at 1,203.69 million[302]. Nokia has been present in India for a considerable period, with its Chennai factory (South India) being the first to

---

[299] https://www.bbc.co.uk/news/world-asia-india-66594520
[300] https://www.cbc.ca/news/science/india-fourth-country-to-land-on-moon-1.6944716
[301] https://www.gsma.com/about-us/regions/asia-pacific/wp-content/uploads/2022/09/India-report-FINAL-WEB.pdf
[302] https://www.ibef.org/industry/indian-telecommunications-industry-analysis-presentation

produce 5G radio equipment in the country[303]: Nokia's state-of-the-art manufacturing unit started manufacturing 5G New Radio (NR) based on the 3GPP 5G New Radio Release 15 standard in 2018. This factory also manufactures 2G, 3G, and 4G units and serves both domestic and international markets. The Indian Government has also implemented a Production-Linked Incentive (PLI) Scheme to boost domestic manufacturing of telecom and networking products. As of December 2022, 42 companies had committed to investing Rs. 4,115 crore (US$ 502.95 million) under this scheme[304].

I believe India uniquely has all the motivations and ingredients to address all the supply and demand side challenges of bridging the digital divide – and even driving the manufacturing of cheaper mobile technology standards. Expensive 5G standards and equipment are not the answer for rural India. India knows this well.

### 7.4.2  India: Towards a *Pro-unconnected* Mobile Standard

*Consider a thought experiment…*

What if India, in collaboration with other lower-income countries (LICs), lower-middle-income countries (LMICs), least developed countries (LDCs), or even an upper-middle-income country (UMIC) like Brazil, developed a revised *'4G/LTE mobile standard for rural connectivity'* with the following five core principles or tenets? [I acknowledge Prof. William Webb[305] who added to a couple of the tenets below]:

1.  **Starting with the 4G/LTE standard**: this wireless communication standard improves on 3G and 2G with higher speeds, lower latencies, greater capacity, and more efficient radio technology. 4G LTE has revolutionised mobile communications, forming the basis of today's connected data world.

---

[303] https://www.nokia.com/about-us/news/releases/2018/10/25/nokia-chennai-factory-first-to-manufacture-5g-radio-equipment-in-india/
[304] https://www.ibef.org/industry/indian-telecommunications-industry-analysis-presentation
[305] https://scholar.google.co.uk/citations?user=wAHZ158AAAAJ&hl=en

2. **Remove all the Voice over LTE (VoLTE) standards from it**: – ensuring that the new 4G remains just all about data, and not voice. It is true that using LTE networks for voice, VoLTE eliminates the need for calls to switch back to 2G or 3G networks, and that 2G and 3G networks are gradually phased out globally[306]. However, the VoLTE stack of standards increases the complexity and cost of the 4G LTE standard. I acknowledge that removing this stack would simplify the new residual standard, limiting it to support only WhatsApp-type voice calls, but this already suffices for most LDCs I know today – particularly with the fall back to 2G.

3. **Add seamless Wi-Fi integration:** as Wi-Fi is so key to connecting the 2.6 billion unconnected, this makes much sense to me. Seamless Wi-Fi integration with 4G/LTE would bring a whole host of benefits and enhance connectivity including:

   a. *Improved coverage*: Switching between 4G/LTE and Wi-Fi networks help maintain connectivity in areas with poor cellular coverage, such as indoors.
   b. *Cost Savings*: Redirecting data traffic to Wi-Fi networks can alleviate the load on mobile or cellular networks and significantly reduce data expenses for users. This is particularly beneficial in regions where mobile data plans are limited or costly.
   c. *Improved Performance*: Integrating the capabilities of 4G LTE and Wi-Fi networks can result in enhanced performance, reduced latency, and increased data speeds, particularly in crowded areas.
   d. *Enhanced User* Experience: Users can enjoy a more reliable and consistent connection, whether they're browsing the internet, streaming videos, or making VoIP calls.
   e. *Capacity Management:* Operators can manage network capacity by routing traffic between mobile/cellular and Wi-Fi networks, optimizing resources, and improving network efficiency.

---

[306] It would indeed to controversial stating 2G and 3G networks should not be phased out. I am not so sure it should be that controversial in the context of connecting the billions of unconnected to voice and low-capacity data.

Technologies such as Voice over Wi-Fi (VoWiFi) and Wi-Fi Calling integrate cellular and Wi-Fi networks, enabling users to make voice calls and send messages over Wi-Fi when available.

i.  **Add seamless roaming**[307]: adding seamless roaming to 4G standards for rural connectivity can be a significant change for the following reasons (Popescu *et al.*, 2011):

    a.  *Connecting more users*: users should be able to access any available mobile network.

    b.  *Improved coverage & enhanced user experience:* seamless roaming lets users switch networks without interruption, ensuring continuous connectivity in areas with limited single-provider coverage. Users can benefit from a more reliable and stable connection, which decreases the likelihood of dropped calls and enhances data access in rural areas.

    c.  *Network efficiency and cost savings*: Seamless roaming can optimize network resources by balancing the load across multiple networks, leading to improved overall efficiency. Utilizing existing infrastructure from multiple providers can reduce the cost of expanding coverage in rural areas.

Seamless roaming has the potential to greatly improve rural connectivity by providing enhanced access to digital services and contributing to the reduction of the digital divide.

ii.  **Include and prioritise sub-1GHz spectrums with the revised standard**: prioritising low sub-1GHz spectrums for a revised 4G/LTE standard can provide major benefits, particularly for enhancing rural connectivity and addressing the digital divide:

---

[307] I acknowledge that implementing seamless roaming requires addressing technical challenges such as interoperability, handover mechanisms and network compatibility, making it non-trivial. Effective implementation of seamless roaming requires collaboration among telecom operators, who must establish agreements supported by industry standards and regulatory frameworks.

a. *Improved Coverage*: Low-frequency bands, specifically those below 1GHz (such as 600MHz, 700MHz, 800MHz, and 900MHz), exhibit superior propagation characteristics due to basic physics[308] [309]. These frequencies cover larger areas and penetrate buildings more effectively, making them particularly suitable for rural and remote regions.

b. *Cost effective networks:* deploying networks utilising low-frequency bands would result in cost savings, as fewer base stations are required to achieve the same coverage area compared to higher-frequency bands.

c. *Improved indoor coverage*: low-frequency signals penetrate walls and obstacles better, enhancing indoor coverage and user experience.

d. *Efficient spectrum use & IoT/M2M support:* low-frequency bands can optimize spectrum resources, ensuring coverage and capacity in both urban and rural areas. These bands are also suitable for Internet of Things (IoT) and Machine-to-Machine (M2M) communications, which often require extensive coverage and reliable connectivity.

By prioritizing low sub-1GHz spectrums, telecom operators and regulators can enhance connectivity in global underserved areas of India, Africa, South East Asia and Latin America – making it a valuable strategy for bridging the global digital divide.

*End of thought experiment.*

My intention is *not* to persuade the reader to agree with my revised '4G/LTE for rural connectivity standard' outlined above. The main point is that it is a 'standard' based on the requirements and needs of the 2.6 billion people who are currently unconnected to the Internet, as well as the 1,294 – 2,000 million individuals who face challenges in making basic calls regularly (the usage gap problem), as of early 2025.

---

[308] https://gsacom.com/paper/sub-1-ghz-spectrum-for-lte-and-5g/
[309] https://www.gsma.com/connectivity-for-good/spectrum/wp-content/uploads/2021/04/WRC-23-Low-Band-Capacity.pdf

Just contrast my revised 4G/LTE proposition above with the priorities of 5G and what the mobile industry is planning for 6G. Which of them is more relevant to the billions of digitally unconnected or under-connected? The reality is that this poor tail 2.6 billion unconnected and the [minimum] of 1.294 billion who cannot make a basic voice call do *not* have the economic power to drive future mobile technology standards.

However, a country with the population, rural connectivity experience to date and technology expertise of India can certainly lead a joint initiative with other countries to 'intervene' in the realisation of a more fit for purpose standard – one that better addresses the needs of the billions unconnected and under-connected.

### 7.4.3 A Potential Indian-Led Initiative: More Than Just Tech

An Indian-led joint initiative can lead on more than just technology. I think an Indian-led digital divide joint initiative [with selected LDCs or EDCs or even UMICs] can lead on more than just technology standards that I elaborate on above. India is also well-positioned to lead on multiple fronts as regards addressing the digital divide.

1.  *Research and Innovation*: India's deep relationship with mobile technology heavyweight, Nokia, and its growing tech ecosystem [including startups and research institutions], can contribute to developing new technologies and solutions that address the digital divide. Joint initiative, collaborative research projects and knowledge exchange can benefit multiple countries – just like joint initiative Horizon R&D projects benefit the EU.

2.  *Digital divide affordable solutions:* After India's successful Chandrayaan-3 lunar mission, I know she has the potential to *jointly lead* and evolve – not only cheaper mobile standards – but to lead on innovations in low-cost technology and connectivity solutions that can be shared with other nations. This includes affordable smartphones and innovative rural connectivity solutions.

3.  *Policy and Regulation*: India's proven experience with digital policies and regulations, such as the Digital India initiative and the BharatNet programme, can serve as a model for other LIC countries. Sharing best

practices and frameworks can help other nations develop their own strategies to bridge the digital divide.

4. *Capacity Building*: an Indian-led joint initiative would provide truly relevant training and capacity-building programs for other LDC-type countries. India truly feels their digital divide pains, which most developed nations really do not. This includes developing digital skills for SMEs, schools, NGOs and consumers, promoting digital literacy, and creating educational resources.

5. *Public-Private Partnerships*: The BharatNet programme described earlier demonstrates many examples of India's successful public-private partnerships in order to address the digital divide – and they could be replicated in other countries. Fostering collaboration between Government, States, countries, industry, and civil society, an Indian-led joint initiative could create sustainable solutions to the digital divide.

These reasons are not exhaustive, but they suffice to demonstrate why an Indian-led joint initiative to address the digital divide globally is compelling.

## 7.5 Summary

This chapter seeks to examine the key components of individual country strategies as well as approaches to inter-country collaboration that could enhance connectivity among the estimated 2.6 billion people not connected to the Internet as of early 2025, while also facilitating more meaningful access for up to 2 billion additional individuals. Furthermore, the chapter outlines the potential contributions that a major economy such as India could make toward these objectives.

# 8    Policy & Regulation: Towards Mitigating the Digital Divide

In Chapter 1 of Nwana (2024), I argue passionately that policy and regulation is the fifth – but arguably the most important – driver of any sector of the Economy. I argue that policy and regulation truly help 'wield together' from the other four drivers of every sector, namely macroeconomy, technology, industry (supply side) and consumers/citizens (demand side). Policy and regulation – when done properly – weaves together the latter four drivers to enable a healthy, growing and thriving sector full of competition, maximum inclusion, low prices, more consumer choice, safer services, innovation, etc.

I genuinely do not believe that the 'lazy' ICT Ministers, policy makers and regulators of many LDCs, LLDCs, LICS, LMICs and SIDS – that I call out in the previous chapter [cf. Section 7.3.2] – truly get the above first paragraph of this chapter. None of these individuals should be in their positions if they do not fully understand the centrality of policy and regulation in making any sectors of their economies work efficiently. This includes ministers, policymakers and regulators of other network industry sectors such as electricity, water and waste water, transportation, aviation, rail, postal, gas, banking, etc. I am so enthusiastic about good regulation that it led me to write a 450+ page book on Demystifying Economic Regulation (Nwana, 2024). I am also equally enthusiastic about Telecommunications, Media & Technology (TMT) and Digital Economy policies that I authored two books about these areas (Nwana, 2014, 2022). So, I do not intend to repeat the treatises of these three books in this chapter.

However, and rather, I am much briefer, and I concentrate my commentary on connectivity digital divide policy and regulation, where possible drawing from other authoritative and/or seminal papers and reports from reputable organisations like the ITU, the Broadband Commission, the GSMA and others.

## 8.1 GSMA: Driving Digitalisation Through Policy & Regulation

In its many reports that the GSMA publishes annually, it usually emphasises the centrality of policy and regulation to address digitalisation. Therefore, permit me to reiterate the importance of policy and regulation to addressing digitalisation and digitisation issues by drawing from a recent report from the GSMA. I would like sector ministers, regulators and other key ICT policy leaders from LDC, LIC, LMIC, SIDS, LLDC and EDC nations to fully understand this.

During 2024, GSMA Intelligence[310] published their first [seminal] report on digitalisation in Africa entitled *Digitalisation and the Africa We Want – Introducing the GSMA Digital Africa Index*[311]. Though the report is all about Africa, and all about digitalisation [not necessarily only about the digital divide], I believe its core messages apply to all of the nations that this book is aimed at, as regards their digital divide challenges: i.e., LDCs, LICs, LMICs, EDCs, SIDS, LLDCs and even some UMICs.

A core message of the report is the following: in 41 countries reviewed and indexed in Africa, *digital development is significantly hampered by weak policy and regulations*. More specifically pertaining to the digital divide across Africa, the report is categorical on Africa's chances of connecting all to the Internet by 2030:

> "while the transformative socioeconomic impacts of digital technologies are well established, a digital divide persists in Africa, where around two thirds of the population do *not* currently use mobile internet. *Based on recent connectivity trends, it will take another 30 years for the usage gap in mobile internet connectivity to close in Africa"*[312] (*my emphases*).

---

[310] The GSMA self-describes GSMA Intelligence as "GSMA Intelligence is the definitive source of global mobile operator data, analysis and forecasts, and publisher of authoritative industry reports and research. Our data covers every operator group, network and MVNO in every country worldwide – from Afghanistan to Zimbabwe. It is the most accurate and complete set of industry metrics available, comprising tens of millions of individual data points, updated daily… Our team of analysts and experts produce regular thought-leading research reports across a range of industry topics". Source: see link below.

[311] https://data.gsmaintelligence.com/api-web/v2/research-file-download?id=88244627&file=221024-Digital-Africa-Index-EN.pdf

[312] *Ibid.*, p.5

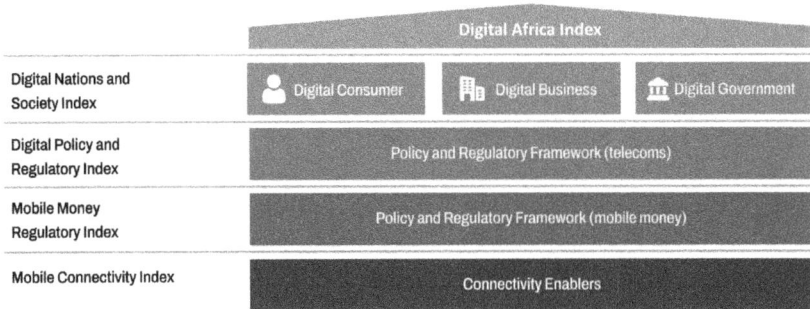

| Digital Nations and Society Index | Digital Consumer | Digital Business | Digital Government |
| Digital Policy and Regulatory Index | Policy and Regulatory Framework (telecoms) | | |
| Mobile Money Regulatory Index | Policy and Regulatory Framework (mobile money) | | |
| Mobile Connectivity Index | Connectivity Enablers | | |

Figure 8.1 – Digital Africa Index (DAI)
Source: Adapted from GSMA Intelligence 2024[313]

Here are some core takeaways and [new] tools from this GSMA Intelligence report[314]:

i.  As noted earlier, of 41 African countries reviewed and indexed, digital developments are significantly hampered in them by weak policy and regulations.

ii.  The report introduces the GSMA Intelligence's web tool which consists of two composite indices: *the Digital Nations and Society Index* (DNSI) and the *Digital Policy and Regulatory Index* (DPRI)[315]. The web tool provides overall index scores as well as the underlying score for each indicator and a market comparison tool. These two new indices complement two existing GSMA indices namely, the *Mobile Connectivity Index*[316] and the *Mobile Money Regulatory Index*[317]. These two existing indices provide greater insights into how to realise digital and financial inclusion goals in the African region, respectively. All four indices lead to a comprehensive 'Digital Africa Index' as shown in Figure 8.1.

---

[313] https://data.gsmaintelligence.com/api-web/v2/research-file-download?id=88244627&file=221024-Digital-Africa-Index-EN.pdf

[314] *Ibid.*

[315] Both indices are available at the following link: https://www.gsma.com/digital-africa-index.

[316] https://www.mobileconnectivityindex.com/

[317] https://www.gsma.com/mobilemoneymetrics/#regulatory-index

iii.     According to the first set of findings of the GSMA's Digital Africa Index (DAI), the 41 African countries surveyed are currently experiencing low digital development *due to unfavourable policy and regulatory frameworks* (see Figure 8.1). These countries include the one I hail from – Cameroon – along with others such as Algeria, Tunisia, Ghana and Cabo Verde. The report classifies these 41 countries as "low digital development with less enabling policy and regulation", noting that "they have significant scope to accelerate digital transformation <u>with more enabling policies and regulations</u>"[318]. Only four countries are rated 'High digital development, more enabling policy and regulation', including South Africa, Mauritius, Kenya and Seychelles – all with Digital Policy and Regulation Index (DPRI) scores above 50. The DPRI analyses the policy and regulatory framework which influence digitisation adoption, and aims to identify the constraints to digital development in Africa. This includes policy and regulatory gaps required to foster inclusive and sustainable digital transformation. The second index, the Digital Nations and Societies Index (DNSI) index, measures digital adoption and usage by consumers, businesses and Governments across the continent of Africa.

iv.      The general weak policy and regulatory environments also impact the digital developments of Governments, corporates, SMEs and NGOs, e.g. e-Government, increasing the use of P2G and G2P payments, and scaling up Govtech solutions. The use of e-commerce digital technologies (e.g., electronic merchant payments, online transactions, etc.) are largely non-existent beyond several African countries including Kenya, South Africa, Mauritius, Zimbabwe and Namibia. For example, 40 percent of Kenyan SMEs use digital payments[319].

Most countries score low in licensing and spectrum management due to the lack of a spectrum roadmap in many of these countries, not enough IMT spectrum having been assigned, high spectrum fees, and the lack of technology- or service-neutral licenses. GSMA also recommends removing sector-specific taxes on handsets/devices and [mobile] services.

---

[318] *Ibid.*
[319] https://itweb.africa/content/o1Jr5qxPzppqKdWL

The GSMA Intelligence analyses concludes that Africa's GDP could be boosted by circa USD $700 billion over the 2024–2030 period if the continent's *usage gap* is closed by 2030. However, to realise this, the GSMA has identified several *policy* and *regulatory* priorities and blockages that countries must address now.

v.   **Identified policy and regulatory blockages**: the GSMA Report argues that "most countries in Africa have the opportunity to unlock their potential with better policies", and proceeds to identify its recommended key policy and regulatory blockages including the following non-exhaustive list:

   a.   *the lack of monitoring of digital divide and broadband strategies* which clearly measures and evaluates progress made in terms of skills, connectivity among disadvantaged population groups (particularly women and rural populations) and financial accessibility.

   b.   *the lack of efficient (i.e., effective and transparent) universal access service funds* – the report strongly argues for reforming Universal Service Funds (USFs) to ensure they are more efficient, transparent and effective. It also argues for more sustainable delivery of projects to deploy infrastructure and increase connectivity in remote and underserved regions; and

   c.   *the lack of a policy and regulatory framework* favourable for the emergence of technological start-ups.

I humbly submit that these [above five] key takeaways from this GSMA intelligence Report – albeit about digitisation across Africa in 2024 – all apply to broader digital divide challenges across other disadvantaged regions beyond Africa, including the Caribbean, South East Asia and parts of Latin America.

The key takeaway message of this section [from the GSMA] is that policy and regulation is key to addressing the digital divide usage and coverage gaps challenges.

## 8.2 Government Digital Divide Policy

The previous section presents the evidence-based case that digital development across LICs and LMICs [across Africa] are significantly hampered *by weak and unfavourable policy and regulatory frameworks*. Ergo – the connectivity digital divide in these countries and elsewhere requires strong and favourable policy and regulatory frameworks. So, what does a strong and favourable policy look like?

Permit me to start responding to this question of Government policy on digital divide connectivity by reverting to the case study of India. Recall some key tenets of India's BharatNet connectivity programme described in the previous chapter:

- BharatNet is arguably the world's largest rural connectivity initiative – an initiative set out clearly in a long-term priority Government digital divide (rural connectivity) policy.
- Recall too that in early 2010, the Internet penetration in India was merely 92 million users compared to some 900M[320] today in 2025. India instituted a major rural connectivity Government policy to address this market failure. The policy also decided to fund the programme from the Indian Universal Service Obligation Fund (USOF) – a major regulatory intervention – administered by the Department of Telecommunication (DoT)[321][322].
- Government policy ensures that the Digital India policy initiative aims (and continues) to provide people-centric services, including for marginalised groups too. This has led to the flagship 'Accessible India Campaign and mobile app' which facilitates realising universal accessibility, including Indians with disabilities to have access to equal digital opportunities[323]. By September 2024, the campaign had trained over 1,250 sign language interpreters, with 588 State and 95 Central Government websites accessible for persons with disabilities.

---

[320] "Internet users by region and country, 2010-2016." Accessed: Feb. 10, 2025. [Online]. Available: https://www.itu.int/en/ITU-D/Statistics/Pages/stat/Treemap.aspx
[321] "USOF Ongoing Schemes - Advancing Telecom Connectivity in India." Accessed: Feb. 10, 2025. [Online]. Available: https://usof.gov.in/en/ongoing-schemes
[322] "BharatNet: Bridging the Digital Divide." Accessed: Feb. 10, 2025. [Online]. Available: https://pib.gov.in/PressReleseDetailm.aspx?PRID=2086701&reg=3&lang=1
[323] https://www.broadbandcommission.org/publication/state-of-broadband-2024/, p.8

- Government policy also ensured that the entire programme is implemented and managed under Bharat Broadband Network Limited (BBNL), which is responsible for planning, deploying and managing the BharatNet project.

My key message of this section thus far is this: at the driving core of the BharatNet programme has been a *sustained Government rural connectivity policy* which dates back from 2010 to date (in 2025). Furthermore, the policy should not only be 'living', but it should also have clear advocacy targets.

### 8.2.1 SMART Government Digital Divide Policies with Clear Advocacy Targets

Government digital divide policies must be SMART (i.e., Specific, Measurable, Actionable, Realistic and Time-Bound), with clear Advocacy Targets, as the Indian BharatNet programme demonstrates. They must be regularly measured and evaluated, at least annually. Indeed, from the perspective of supply- and demand-side digital divide challenges, I strongly recommend that your country's *Digital Divide (or Rural Connectivity) Policy* adopts the recommended Broadband Commission Advocacy Targets[324]. The targets assess progress in broadband access and identify remaining gaps. However, recall that in addition to the tail 2.6 billion unconnected to the Internet, I have argued [and established] that there is a minimum of 1.294 billion (see Table 6.1) who cannot even effect a basic voice call. For this reason, I offer some *modifications or enhancements*[325] to some of the Broadband Commission's advocacy targets:

- **Advocacy Target 1**: *Make broadband policy universal*. I demure slightly with the Broadband Commission here and modify this target to *'Make digital divide policy universal'*. As part of this target, the Broadband Commission recommends the developing and implementing of National Broadband Plans (NBPs) as critical to advancing connectivity. Again, I demur slightly, as I argue in the last chapter that Broadband plans should

---

[324] https://www.broadbandcommission.org/publication/state-of-broadband-2024/, Chapter 2.
[325] Recall this is because, in addition to the tail 2.6 billion unconnected to the Internet, I have established that there is a minimum of 1.294 billion (see Table 6.1) who cannot make a basic voice call.

evolve into much broader '*Mitigating Digital Divide Plans'* (cf. Section 7.1.2).

- **Advocacy Target 2**: *Make broadband more affordable*. I enhance this target to read '*Make digital voice and broadband services more available'*. The Commission rightly argues that affordability of [broadband[326]] services remains a significant barrier to universal digital connectivity. So, this Advocacy Target specifies that entry-level broadband services should be made more affordable in LMICs by 2025 (i.e. this year), the rationale being that this is key to achieving both universal and meaningful connectivity. This target specifies that, in developing countries by 2025, entry-level [broadband[327]] services prices should be below 2% of monthly GNI per capita. This target advocates that both the Total Cost of Ownership (TCO) and use of [broadband] devices and connectivity are considered: handset affordability and availability, device affordability, reliability and durability and device distribution. This target also emphasises the addressing of demand-side issues such as digital illiteracy and lack of awareness of using the Internet.

- **Advocacy Target 3**: *Get everyone online*: I enhance this target to read '*Achieve Universal voice and broadband connectivity'*. Achieving universal digital connectivity of both voice and data are clearly pressing goals because of the 2.6 billion offline and the 1.294 billion usage gap for voice. It is crucial to bridge these connectivity gaps, particularly in remote and underserved areas. The Broadband Commission believes that emerging technologies like satellite and direct-to-device (D2D) offer promising solutions to bridge these coverage gaps. However, as I warn in Chapter 6 [Section 6.3], beware of the hype and promises of [new] sexy technology solutions. Besides, they do not even address the usage gap challenges.

- **Advocacy Target 4**: *Promote Digital skills development*. I have no enhancement to this advocacy target. Digital literacy is a critical demand-side challenge that I emphasise often in this book, with billions lacking the skills needed to effectively use the Internet. This Advocacy Target calls for 60% of adults and youths to have achieved [at least] a minimum level of

---

[326] and to which I add voice too
[327] *Ibid.*

proficiency in sustainable digital skills by 2025. Developing these skills is essential for them to fully participate in the digital economy and leverage its benefits.

- **Advocacy Target 5**: *Increase use of digital financial services*: I have no enhancement to this advocacy target either. Digital financial services, like M-Pesa (cf. Section 5.1), offer tremendous opportunities to rapidly increase citizens using the Internet, whilst extending access to its social and economic benefits – enabling them to participate in the growing digital economy. This target aims to have 40% of the world's population using digital financial services by 2025. "In 2018, 2 billion adults did *not* have access to a bank account, and yet 1.6 billion adults had access to a mobile phone, creating the potential for e-finance access, and with this, access to economic empowerment"[328]. However, significant [mostly] demand-side barriers mitigate this target from being achieved: digital literacy, poverty, economic constraints, and trust issues. Addressing these challenges is crucial to harness the full potential of the digital economy and digital financial services (Nwana, 2022).

- **Advocacy Target 6**: *Get MSMEs or SMEs online:* I have no enhancement to this advocacy target either. Micro-, Small-, and Medium-sized Enterprises (MSMEs) are the vital backbones to all LMIC and LDC nations, as with developed HIC nations too. The UN Broadband Commission's *ambitious* Advocacy Target 6 focuses on improving the connectivity to the Internet of MSMEs by 50% over the period 2018 to 2025. It is ambitious, not least because the 2024 Broadband Commission report itself notes that:

> "an IFC/World Bank survey of 3,325 microenterprises in seven African countries found low levels of smartphone and computer use. Use of the Internet for business purposes was *7% on average*, ranging from 24% in South Africa to *1% in Rwanda. Computer ownership is also low with over 90% of businesses surveyed in Ghana, Kenya, Mozambique, Nigeria, Rwanda, Tanzania and Uganda reporting not having one.* Most cited not having a need for Internet access or computers in their business. A UNDP survey

---

[328]https://www.broadbandcommission.org/publication/state-of-broadband-2024/, p.42

focusing on MSMEs in Kenya revealed that they were adversely affected by the pandemic, with one out of every 10 enterprises surveyed indicating a shutdown of their businesses due to the pandemic[329]" (*my emphases*).

This is exactly my experience across most LDC, LIC and LMIC nations that I have worked in. However, this advocacy target remains key. Bridging this digital divide is essential to empower local SMEs/MSMEs to compete in global markets.

- **Advocacy Target 7**: *Bridge the gender digital divide:* I also have no enhancement to this advocacy target. This target states that, by 2025, gender equality should be achieved across *all the other targets*. Gender disparities in Internet usage are very pronounced in LMICs, LICs, SIDS, LLDCs, and LDCs in general. Addressing these disparities is crucial for ensuring equitable access to digital opportunities. Countries must focus on promoting universal digital inclusion and empowering women and the disabled to participate fully in the digital economy.

These [modified] advocacy targets are all truly key to any good government policy towards bridging the Digital Divide in most, if not all, LDCs.

### 8.2.2 From ITU Policy to National Digital Divide Policy: Adaptations

National digital divide policy would benefit from adaptations (or 'translations') from ITU policy. This is because (as I hope the reader would now agree) that it is *one thing to* have clear multinational agency-led connectivity policy, e.g. the ITU Broadband Commission's above (Section 8.2.1), along with its recommended seven advocacy targets of the previous section. However, it is *another thing* to 'translate' these advocacy-led policy 'targets' and principles into a local national context like PNG (cf. Section 2.2) or Malawi (cf. Section 2.3).

It is obviously clear [indeed, I have asserted in Chapter 2 that it was clear back in 2018] that Malawi and PNG would never meet the seven advocacy targets of the previous section by 2025 [i.e., this year]. For example, reflect on the UN/ITU

---

[329] *Ibid.,* p.45

Broadband Commission's ambitious Advocacy Target 6 – on improving the connectivity to the Internet of SMEs by 50% over the period 2018 to 2025. Consider the reality that an IFC/World Bank 2024 survey of 3,325 microenterprises in seven African countries found: they found that the use of the Internet for business purposes was 7% on average, ranging from 24% in South Africa to 1% in Rwanda.

Therefore, it appears to me that LDC, LLDC, LMIC, LIC or SIDS nations should set more relevant, contextual SMART targets for their local countries, for example, *not expecting South Africa and Rwanda to realise the same target by the same date* – not least because they are starting from very different opening positions when the targets are/were being set. Hence, in the developing and implementing of national Digital Divide Government policies and/or National Broadband Plans (NBPs) [or National Digital Divide Mitigation Plans], it is crucial to <u>adapt</u> all multinational agency-led (e.g. the ITU) policies and advocacy targets into more 'realistic' local ones. Afterall, the 'R' in SMART stands for 'Realistic'.

In prior chapters, I highlight some [supplementary] considerations or recommendations towards devising national digital divide connectivity policy including:

1. *There are stretch digital inclusion targets vs. patently impossible ones* (cf. Section 2.1.1). LDCs, LICs, LIMCs, SIDS, LLDCs and even EDCs need to set more local context, measurable realistic targets.

2. *The implausible 2018 Broadband Commission Targets for 2025* (cf. Section 2.1.2). Perhaps LDCs, LICs, LIMCs, SIDS, LLDCs and even EDCs <u>should be</u> the key nations to influence and set Broadband Commission target dates.

3. *2 billion of those 'connected' to the Internet are not meaningfully connected (cf. Section 6.1.2)*: I believe this passionately, and LDCs, LICs, LIMCs, SIDS, LLDCs and even EDCs should take this most seriously as a key consideration in the setting of their national connectivity policies and plans. I recommend they also engage the ITU to review their definition of 'meaningful' connectivity.

4. *Coverage gaps are one thing, but usage gaps are the much harder aspect to the digital inclusion challenge* (cf. Section 2.5): national digital divide

policies should acknowledge this difficult reality, and develop commensurate strategies and digital divide mitigation plans.

5. *Intra-Country strategy, implementation and collaboration plan for widescale rural connectivity* (cf. Section 7.2): the policy must be accompanied by such a strategy, implementation and collaboration plan. This is because of the high risks of intra-country coordination failures.

6. *Inter-Countries Collaboration strategies for mitigating the digital divide* (cf. Section 7.3): after a country, let us say Nepal, develops its digital divide connectivity policy, she may choose to seek a collaboration with its neighbour India – on both countries working together to address their respective digital divide challenges, sharing best practices, standards, technology solutions, capacity building, etc.

7. *The massive 'Load Effort' needed to 'lift' millions in most LDCs into digital voice/data connectivity demands a redefinition and rethinking of broadband connectivity as a 'universal service'* (cf. Section 3.4): I recommend national policies seriously take this recommendation into consideration.

8. *The digital divide Load Effort also demands empowering a 'bottom-up' 'community-led' ownership mindset* (cf. Section 3.5): just like what the BharatNet rural connectivity programme in India is driving to realise.

9. *The digital divide load effort also demands Collaboration & Cooperation amongst national Broadband Stakeholders*: recall, the BharatNet programme employs multiple implementation models, including State-led, public sector/CPSU, private sector, and satellite connectivity stakeholders and approaches. Other LDCs should learn from this (cf. Section 7.2).

10. *Democratising and localising connectivity responsibility of the 2.6 billion not online* (cf. Section 5.3): this follows on from the former. I believe this is a core tenet of any national digital divide connectivity/rural connectivity policy.

11. *The telecoms market has failed key [rural] populations – with a huge usage gap of 1.3 billion [minimum] not using basic 2G voice services* (cf. Section 5.1): most LDCs I know still suffer from significant basic voice market failures, less so due to coverage gaps – but more due to usage gap failures. Any national digital divide connectivity/rural connectivity policy must prioritise this challenge. Basic voice connectivity is surely a more pressing 'utility'.

This list of digital divide policy considerations above is not exhaustive vis-à-vis the preceding chapters [to this one] in this book. They are meant to encourage a more comprehensive and exhaustive rethinking of how national connectivity policies, strategies and plans are developed and/or evolved.

## 8.3  Complementing Digital Divide Policy with <u>Firm Regulation</u>

A strong and favourable Government digital divide *policy* obviously requires an equally strong, favourable and complementary *regulatory* component – to realise a strong and favourable policy and regulatory framework (cf. Section 8.1). Indeed, the role of *regulation* is most critical.

Recall that in Chapter 6, I make the [hopefully strong] case that *regulatory [and policy] interventions* are very necessary with addressing the digital divide and rural connectivity challenge *(cf. Section 6.2)*. I argue that, as depicted in Figure 6.4, to seek to achieve positive outcomes for consumers and citizens that do not arise naturally from market *competition* [on the right of Figure 6.4], i.e. a market failure has arisen – *regulation* [on the left of the same figure] becomes necessary to bring such net benefits. I make the case that most of the 2.6 billion unconnected to the Internet (as of 2025) will *not* be connected via the market, i.e. through competition. So, I insist that regulatory interventions – as is happening with the BharatNet programme in India with their USOF – is certainly a key aspect of the way forward to address the unconnected tail 2.6 billion.

### 8.3.1  Listing Methods of the Practice of Regulation

In Nwana (2024, Chapter 4), I overview in detail many methods of the practice of regulation. In the above chapter of this 2024 book, I note that the methods include licensing and certification, access (non-price) regulation, access price regulation, rate of return and price cap/revenue cap [price controls] regulation, SMP and *relevant markets competition regulation* (including promoting new entrants), consumer protection regulation, quality and safety regulation, scarce resources & environmental/climate regulation, *universal service/access regulation*, enforcement & responsive regulation, risk-based regulation, accounting/functional/structural separation, taxation and subsidies, impact assessments and costs benefits analyses.

There are two reasons why I enumerate these methods of the practice of regulation:

i. The challenge of connecting most of the 2.6 billion unconnected today in 2025; and
ii. The challenge of improving the meaningful connectivity of circa 2 billion connected today.

I would like the reader to pause and reflect on these two challenges, and quickly realise that different primary methods of the practice of regulation lend themselves to addressing them.

Consider the first challenge: connecting most of the 2.6 billion individuals who are currently unconnected by 2025. While various regulatory methods mentioned earlier will be utilised by authorities in their efforts to reduce the digital divide within their nations, the primary regulatory methods most pertinent to addressing the 2.6 billion challenge include S*ignificant Market Power (SMP) regulation and relevant markets competition regulation* (such as promoting new entrants) as well as *universal service/access regulation.* These methods are emphasised in italics for their importance in this context.

To address the second challenge of improving the meaningful connectivity of approximately 2 billion people connected today, the primary relevant method of regulation to employ is *SMP and relevant markets (competition) regulation, including promoting new entrants.* This approach is relevant because competition from existing market players has already achieved the connection of these 2 billion individuals, though they are *not* meaningfully connected. In addition, *universal service/access regulation* is also an important method to consider for meaningfully connecting the 2 billion people who are not expected to be meaningfully connected by 2025.

I therefore draw from Nwana (2024, Chapter 4) for the rest of this entire section, as I mainly overview universal service/access regulation.

In both [challenge] scenarios, other methods of regulation are also used to complement the primary method(s), including licensing and certification, access (non-price) regulation, access price regulation, rate of return and price cap/revenue cap [price controls] regulation, consumer protection regulation, quality and safety regulation, scarce resources (like spectrum and numbers) regulation, etc. – see Nwana (2024) for details. I think it is important to highlight this distinction of the primary method(s) of regulation that applies to these two challenge scenarios.

## 8.3.2  The Key Role of Wholesale Competition & SMP Regulation

I observe in many LDCs across the globe the increasing failures of their regulators to promote competition, resulting in them relying more on the interventions of increasing regulations. This is truly a pity.

As Figure 6.4 (in Chapter 6) illustrates, competition from licensed operators [on the right of Figure 6.4] is usually the starting point to seek to achieve positive outcomes for consumers and citizens. It shows that if competition is working fine, there is less or no need for regulation. Clearly, competition to date has *not worked* to deliver connectivity for the 2.6 billion unconnected. Nevertheless, this does not mean that a significant proportion of them would not be connected through competition, particularly through better *wholesale competition.*

It is truly key to maximise wholesale competition regulation. Wholesale competition/Significant Market Power (SMP) regulation remain vital regulatory methods for connecting the 2.6 billion people who remain unconnected – typically in the following economic relevant markets or their equivalents in LDCs (see Nwana, 2024, Section 4.5):

- *Wholesale mobile broadband access market:* which could enable and facilitate retail coverage and competition from dozens of MVNOs and broadband ISPs in the country;
- *Wholesale fixed broadband access market:* ditto like above, facilitating access to essential fixed broadband facilities, e.g. national fibre backbones which may be controlled by one SMP player;
- *Wholesale fixed local access market (WFLA):* e.g., to promote fair access to competitors to essential fixed [fibre, FWA, copper DSL] connections into

office buildings, shopping malls, office parks, airports, train stations, universities, etc. – that may be controlled by a single SMP players;

- *Wholesale mobile local access market (WMLA):* ditto to the latter, but for mobile networks;
- *Wholesale national market for voice call termination and roaming on public mobile networks:* this is key to enable and promote access to competitor MNOs to consumers of any other SMP public MNO who already has built networks in some rural areas;
- *Wholesale dedicated capacity (WDC) market:* this market typically involves the provision of dedicated data connectivity between fixed locations. SMP WDC players provide a dedicated private data link with fixed bandwidth to other retail competitors. This service setup delivers consistent speeds and reliability, suitable for large businesses or public sector organisations that need secure and uninterrupted data transfers. This could apply to SMP 'first mile' subsea cable operators, access to SMP satellite capacity, access to SMP metro fibre capacity, etc;
- *Wholesale physical infrastructures market:* wholesale access to physical sites infrastructure refers to the ability of telecom operators to use existing physical assets, such as current physical broadcast and MNO sites, ducts, poles, and chambers, etc. to deploy their networks. This access is crucial for reducing costs and accelerating the rollout of high-speed broadband and other communication services. This approach minimises the need for new civil engineering works, making network deployment more efficient and environmentally friendly[330].
- Etc.

These wholesale markets listed above are not exhaustive, but I believe they are representative of the sort of wholesale economic relevant markets that need to be opened up for competition, to bridge the digital divide more and enhance meaningful connectivity for hundreds of millions more. For hundreds or thousands of ISPs to flourish like in Brazil (cf. Section 5.2), this sort of wholesale regulation is necessary. So, in other to promote much-needed competition, such (or similar) wholesale markets need to be defined in most LDCs, SMP/Dominance assessed in these markets and the necessary remedies imposed. The benefits and outcomes of doing these are clear:

---

[330]https://www.berec.europa.eu/sites/default/files/files/document_register_store/2019/6/ BoR_%2819%29_94_BEREC_Report__Access_physical_infrastructure_updated.pdf

- *SMP wholesale competition regulation aims to provide fair access and encourage competition*: SMP regulation ensures that dominant players in the telecommunications sector do not misuse their market power. By mandating fair pricing and access to infrastructure, smaller providers (like hundreds of ISPs) can participate in the market, promoting competition.
- *SMP wholesale competition regulation encourages innovation and cost reduction:* wholesale competition fosters an environment where providers strive to offer improved services at reduced costs. This cost-effectiveness is important for marginalised and low-income populations who are mostly part of the unconnected 2.6 billion.
- *SMP wholesale competition regulation expands infrastructure*: through regulating access to wholesale networks, operators can collaborate effectively and extend connectivity to underserved regions. This approach prevents redundant investments and ensures that services reach remote or rural areas.

All in all, good SMP wholesale competition would help bridge the digital divide. Policies fostering competition and regulation incentivise operators to focus on inclusivity, prioritizing connecting regions where service provision was previously uneconomical or impractical. Suffice to note to the reader that the 'interplay' of competition and regulation is at the core of the market failure theory, as Nwana (2024) emphasises in Chapter 2. Therefore, the intersection between the two ovals in Figure 6.4 also represents the fact that regulation may be needed to facilitate competition.

### 8.3.3 Universal Service Regulation: Key to Expanding Global Connectivity

I posit that Universal service regulation is the <u>single most relevant method</u> of regulation to employ to connect most of the 2.6 billion unconnected and improve the meaningful connectivity of 2 billion others.

*Universal Service/Access Regulation & Universal Connectivity*: bridging the digital divide of both voice and data services (i.e. universal connectivity), by definition, requires the regulatory tool of universal service/access regulation to realise them. Similarly, the United Nations 2030 Agenda's seventeen SDGs (cf.

Section 1.4 in Chapter 1) are all *de facto* universal service goals. This means from the perspective of Universal Service regulation, these goals represent clear "universal service" obligations/objectives and goals that countries should be striving to achieve. Consider SDG Goals 1 to 3 of 'No Poverty', 'Zero Hunger' and 'Good Health and Well-being' respectively. These are "services" that countries have committed to make "universal" in their countries by 2030. So the SDGs are urgent *call for actions* by all countries – developed and developing – in a global partnership to realise them.

> "The Sustainable Development Goals are the blueprint to achieve a better and more sustainable future for all. They address the global challenges we face, including those related to poverty, inequality, climate change, environmental degradation, peace and justice. The 17 Goals are all interconnected, and to leave no one behind, it is important that we achieve them all by 2030"[331].

The key message here is that the SDGs provide yet another level of why Universal Service regulation is *not* only appropriate – but timely – across a whole host of basic services – including digital connectivity. Countries would need to use policy and regulation to implement these SDGs.

Recall that *universal service* is defined in Chapter 1 (cf. Figure 1.4) as any service that the Government expects to be *available, affordable* and *accessible* throughout the population. So, universal service has two key aspects to it as I explain in Section 3.4. First, it means the service must be made available to all $\underline{X}$[332], perhaps within a given area or of the entire population. Second it should be made available at a uniform and affordable price. Therefore, appropriate universal service obligations should be defined for designated services.

### 8.3.1.1    Universal Service Obligations and Examples

A Universal Service definition for a *designated* service should provide a baseline level of service necessary to everyone. Enhanced services and quality premiums should *not* typically be part of the Universal Service Offer (USO). Most countries therefore would need to designate USOs for the services they deem as

---

[331] https://www.un.org/sustainabledevelopment/sustainable-development-goals/
[332] Governments would typically define what "X"s are, e.g. districts, homes, schools, health centres, etc.

"essential" – and regulators and policy makers would have to find ways to implement them whilst defining and interpreting what "universal" may mean for every designated service context.

Some example USOs drawn from the telecoms industry across the world are captured in Table 8.1 – showing the great variations across countries. Even some developing countries like Pakistan are brave enough to include broadband Internet access as part of its USO – which is incredibly challenging for a country like Pakistan. Perhaps this was driven by the SDGs that were ratified in 2015.

| Country | Universal Service Obligation (USO) Policy |
|---------|-------------------------------------------|
| Hong Kong (Source: OFCA website[333]) | The USO specifies that basic services should be made reasonably available to all persons in Hong Kong, at service charges capped by the published tariffs. *The USO mainly covers basic fixed voice telephony services and public payphones.* Broadband or mobile services are *not* covered under the USO, and the coverage of such services is mainly a commercial decision. |
| Ireland (2016/21 ComReg Decision[334]) | We may designate an undertaking, or undertakings, to satisfy any reasonable request to provide, at a fixed location, a connection to the public communications network ("PCN") and a publicly available telephone service over the network connection that allows for originating and receiving of national and international calls. *The connection must be capable of supporting voice, facsimile and data communications at data rates that are sufficient to permit functional internet access ("FIA").* In this Decision, we refer to this collectively as access at a fixed location ("AFL USO"). |
| Pakistan (Ministry of Information Technology, Telecoms Policy 2015) | The Services falling under scope of USF include the following:<br>• Telephone services to local, national, mobile, toll free, premium rate and international numbers…<br>• Access to emergency services (as under voice licenses).<br>• *Broadband Internet access.*<br>• E-mail, fax and other related services. |

[333] https://www.ofca.gov.hk/en - OFCA Hong Kong
[334] https://www.comreg.ie/media/2021/11/RTC-AFL-USO_R-NON-CONFIDENTIAL-5.11.21.pdf

| | |
|---|---|
| | • Telecentres, including the equipment, buildings and other capex and opex associated with the Telecentre itself.<br><br>• Broadband Internet access to support multiple terminals at telecentres at speeds consistent with the size of the concurrent user base |
| UAE<br>(TRA Policy<br>Document,<br>2017) | Access to the same minimum set of services at the same price, regardless of location – "all consumers in permanent dwellings should have access to services which are *capable of delivering basic voice, TV services, and high-speed data packages of at least 10mbps"*. |
| UK (2020[335]) | From March 2020, consumer has right to request *broadband speed of at least at least* 10Mbps (with monthly price capped at nationally regulated tariff) provided the cost of provision does not exceed £3,400 (although a consumer can pay the excess). |

Table 8.1 – Universal Service Obligation Examples in Telecoms[336]

USOs also do *not* also remain stagnant over time as they are reviewed [and added to] as technologies and the economics of service provisions improve to benefit consumers and citizens more. For example, some countries have extended their *voice-only* telecoms universal obligations to include broadband/internet provisions:

- **Finland** – Broadband Internet was introduced and included in the Finnish telecoms USOs for the first time in 2010, with speeds specified at only 1Mbps in 2010, which was later doubled to 2Mbps in 2015[337], and quintupled to 10Mbps in 2021. Finland became the first country in the world to make broadband a legal right for every citizen in 2010[338].

- **Spain** – the 2011 Law on Sustainable Economy included Internet access of 1Mbps into the USO for the first time (Source - OECD[339]).

[335] https://commonslibrary.parliament.uk/research-briefings/cbp-8146/ - The Universal Service Obligation (USO) for Broadband - House of Commons Library (parliament.uk)
[336] With thanks to Dr Charles Jenne for this compilation of telecoms USOs.
[337] https://omdia.tech.informa.com/om031670/finland-country-regulation-overview--2023
[338] https://www.bbc.co.uk/news/10461048 & https://www.loc.gov/item/global-legal-monitor/2010-07-06/finland-legal-right-to-broadband-for-all-citizens/
[339] https://www.oecd-ilibrary.org/science-and-technology/universal-service-policies-in-the-context-of-national-broadband-plans_5k94gz19flq4-en

- **Malta** – in June 2011, the Maltese regulator mandated universal access to "functional internet access" as a universal service at 4Mbps. This has since been increased to 30Mbps download and 1.5Mbps upload in 2021[340].
- **Switzerland** – a broadband connection providing minimum download and upload speeds of 600/100 Kbps is part of the scope of universal service obligations (Source - OECD[341]).
- **Turkey** – Minimum upwards and downwards broadband speeds of 256 Kbps and 512 Kbps are included as part of the scope of universal service obligations (Source - OECD[342]).
- **The 2018 European Electronic Communications (EECC)** Code[343] embraces universal service broadband as a USO across the EU.

### 8.3.1.2    Driving Universal Services with the World Bank "Gaps Model"

It is one thing setting and specifying the terms and conditions of the USO. It is totally another to make them happen in practice and for them to be enjoyed by all citizens in a country. As an example, many – if not most – countries mandate USOs to ensure the adequate, universal and reasonably priced provision of postal services in all regions of the country, [if not voice/broadband telecommunications too]. Well, in many African and Southeast Asian countries, the postal services USO is <u>close to unachievable</u>, yet alone any telecoms ones. This is because of major [macroeconomic] barriers covered earlier, like poverty in developing economies, but also poor leadership, poor and lazy regulators and policy makers, poor roads and more.

The question here then, is what happens beyond securing the USO? How is it made to come about in practice? The answer is *not* straightforward because it depends on the service concerned (e.g., hospitals, primary education, postal,

[340]https://www.mca.org.mt/sites/default/files/Availability%20of%20Broadband%20as%20a%20Universal%20Service%20-%20Decision%20Notice%20-%2022%20Oct%202021.pdf
[341]https://www.oecd-ilibrary.org/science-and-technology/universal-service-policies-in-the-context-of-national-broadband-plans_5k94gz19flq4-en
[342] *Ibid.*
[343] https://digital-strategy.ec.europa.eu/en/policies/eu-electronic-communications-code - Directive 2018/1972

telecoms, etc.), the dynamics of the sector concerned, whether it is a service entirely provided by the public sector (i.e., the Government), the private sector or some combination of both – and more factors including the realities of the challenging macroeconomy and other barriers. The reality in most countries is that the USO service would need to be provided by some combination of both the private and public sectors. This brings the role of a good competent regulator ever more to the fore. This is because the Government would have to start by clarifying the parties involved in the governance of the USO service concerned.

### 8.3.1.3    Implementing Universal Service – the World Bank "Gaps Model"

I am a big fan of the "Gaps" model (Navas-Sabater *et al.,* 2002) that was pioneered by the World Bank in the context of driving access to telecommunications services for all. I am such a fan of this model because I believe the concepts underpinning this classic model applies to other sectors too. It also implicitly depicts how both the private and public sectors can "cooperate" and "coordinate" to deliver a universal service for all. The model is depicted in Figure 8.2.

The 'theory' underpinning the model is that there is a wide array of mechanisms available for telecommunications decision makers to increase access to telecommunications services – necessitating the need of a general analytical framework model. Navas-Sabater *et al.* (2002) proposed this classic framework of Figure 8.2 called the Gaps Analysis model. In essence, the model divides underserved areas into two different segments or "gaps": the *Market Efficiency Gap* and the *Access Gap*. I believe this model translates to other sectors too.

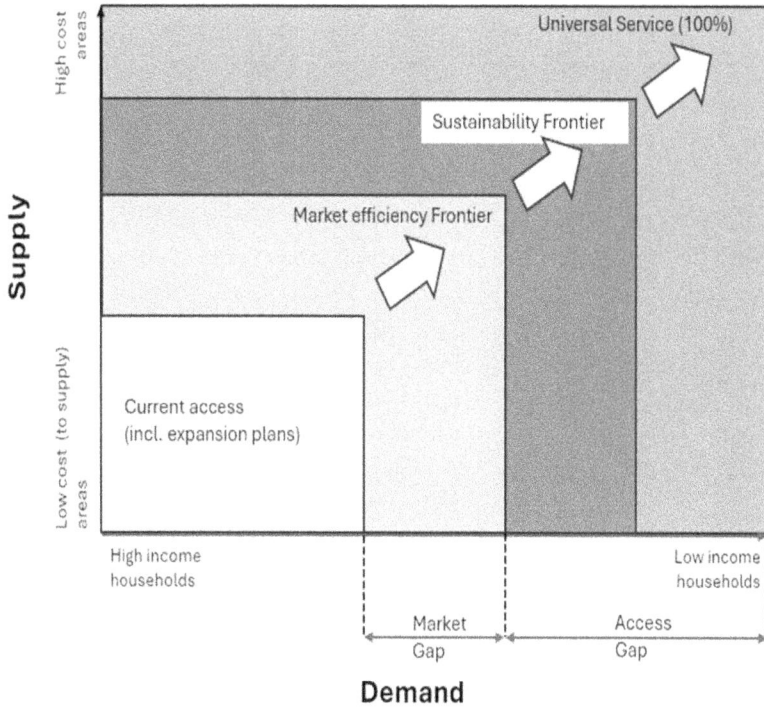

Figure 8.2 – The World Bank Market Gap and Access Gap Model
*Sources: Based on and adapted from Navas-Sabater et al. (2002) & Muente-Kunigami & Navas-Sabater (2010)*

So, let us cover the key tenets of this Market Gap and Access Gap Model:

- **The Axes:** The Y-axis is the *supply-side* axis starting from the lowest costs access supply areas [that commercial players would cover first] to the highest cost access areas [typically rural or least densely populated areas that are more expensive and likely uncommercial to supply]. The X-axis is the demand-side of consumers, citizens and households ranging from the high-income households [or richer citizens who would take up the services first without any hesitation] to lower-income households [or poor citizens who would struggle to pay for any services].

- **The Current Access (including expansion plans)**: this represents the households (or citizens) demand with high-enough incomes to pay and who are currently supplied by services. This area also includes any expansion planned services.

- **The Market Gap & the Market Efficiency Frontier**: The market gap refers to that difference between the level of demand penetration that can be reached under current plans (including expansion plans) and that [typically] *private-sector*-led level that the market *could* achieve by means of an ideal competition law-driven, regulatory and legal environment. This gap could be eliminated with adequate changes in current regulations (in addition to new competitive measures taken by regulators) – and therefore this gap should *not* require public transfers or subsidies. In other words, the application of the optimal set of competition law-driven vs. regulatory instruments for the market sector would close the market gap to the *market efficiency frontier*.

- **The Access Gap:** alternatively, the access gap depicts that portion of the market that even under the optimal legal, competition law-driven and regulatory environment would *not* be reached by market players due to their high costs to service and/or low-income levels. Public subsidies are usually considered to go beyond the coverage beyond the market efficiency frontier into the access gap.

- **The Sustainability Frontier & the Universal Service Frontier**: the Access Gap can itself be split into two different zones as seen in Figure 8.2: the first is the *Sustainability Frontier* and the second is the *Universal Service Frontier* (ideally 100% though it would realistically be less for many designated services in most countries). The sustainability frontier divides two different types of projects: first, those projects requiring only a one-off subsidy to 'kick start' operations and thereafter the areas support themselves on an ongoing basis; second, those projects that would require ongoing subsidies in perpetuity. From an economic regulatory perspective, this is a most important distinction. It is clearly efficient to discern projects that are expected to recover their operational costs and remain profitable from a singular "jump start" public financing, compared to those that will always require public subsidies.

- **The Boundaries of the Market and Access Gap Models vary by country and by service**: it is important to note explicitly – what may be obvious – that all boundaries under this framework are *not* static. Rather, they depend on many factors including all the sector drivers noted at the beginning of this chapter, including technology, macroeconomy, industry, consumer behaviour, innovation of new services/applications, and others. The current access, market efficiency frontier, sustainability frontier and universal service frontier boundaries for a mobile 2G service in a country may be 70%, 80%, 90% and 100%, respectively. However, the boundaries for electricity services in the same country may be 30%, 50%, 60% and 80%, respectively. Note that these latter boundary numbers are much less than for the former and note that a universal service boundary of no more than 80% is proposed for electricity – suggesting that the Government may have [had] to concede that a 100% electricity service in that country is unachievable for the foreseeable future. This sort of reality speaks to a continuous review of universal access targets, policies and mechanisms, and the need to adapt them.

I contend that this 2002-proposed World Bank Market Gap and Access Gap model is still most relevant to good economic regulation of most universal services across most sectors. It forces good regulators to do their utmost to estimate the boundaries for the various services and use them as evidence – amongst other key considerations like the outputs of relevant market reviews. This way, good regulators and/or Governments can:

- Optimise competition law and regulation in order to reach the market efficiency frontier for the service.
- Investigate and implement concrete policies, along with the right choice of projects (and geographical areas) that only require one-off subsidies to 'kick start' operations to reach the sustainability frontier.
- Clearly discern the projects and areas that would always require subsidies in perpetuity.

Hopefully, the reader can see the power of this model shown in Figure 8.2 in the implementation – not only of USO schemes in particular – but also in using good economic regulation to ascertain the boundaries of where regulatory interventions, in terms of one-off or perpetual public subsidies, are required.

### 8.3.4 Essential Components of Effective Digital Divide Regulation

I also recommend that a good digital divide set of regulations and/or framework should comprise several key tenets:

1. *Regulations implementing transparent policies and universal service objectives (USOs)*: the mitigating digital divide regulations must implement clear and transparent mitigating digital divide policies that promote competition among different types of service providers, e.g. big MNOs, hundreds if not thousands of Internet Service Providers (ISPs, as in Brazil), satellite players, etc. As seen from the Brazilian example (Section 5.2), ISPs, particularly smaller, independent, and local operators, play a key role in bridging this divide, and are well-positioned to connect underserved communities due to their local knowledge and presence on the ground. This has been proven to boost economies and improve service quality. As I cover in the previous section, the universal service/access obligations (USOs) or goals must be clearly and transparently set (cf. Table 8.1), recognizing digital connectivity as a basic human right and a transformational force for banking, education, healthcare, and economic growth. Regulations should implement such transparent USOs.

2. *Technology-neutral regulations*: mitigating digital divide regulations should encourage the use of various technologies by ensuring that regulations implemented do *not* favour one technology over another. Such a framework allows for the rapid deployment of innovative services.

3. *Support for new technologies*: following from the latter, no technology (e.g. IMT technologies) should become the *de facto* monopoly technology to mitigate the digital divide in a LDC country. I note this because the regulations and rules that underpin many Universal Service Funds (USFs) in countries I have worked in implicitly favour IMT technologies like 2G/3G/4G over others such as fibre, Wi-Fi, FWA, satellite, TVWS or other innovative connectivity technologies. Fostering the adoption of new technologies, such as Non-Geostationary Satellite Orbit constellations (NGSOs) can provide broadband to remote, rural and unconnected populations.

4. *Collaboration, partnerships and standards-based regulations*: the previous chapter (Chapter 7) emphasises this. Collaboration between governments,

private sector, and non-governmental organisations to address the digital divide is truly key in many contexts – including employing harmonised regulations. Regulations that have resulted in more sustainable and pragmatic solutions in similar contexts should be considered, if not implemented, e.g. regulations (and policies) that encourage community-based networks, public-private partnerships and more such collaborations. Standards-based regulations are clearly key too – why deviate from proven standards in other markets, or from international standard organisations such as the ITU or 3GPP?

5. *Regulatory innovation*: no mitigating digital divide regulatory framework and set of regulations should be stationary and fixed. I recommend they are continuously updated, ensuring there exists an innovative regulatory framework that keeps pace with technological advancements and changing market dynamics.

By incorporating these tenets, ICT regulators can create a supportive environment that helps bridge the digital divide and ensures most citizens have access to the benefits of digital connectivity.

### 8.3.5 Techno-Economic Modelling: A Data-Driven Digital Divide Regulation Tool

Implementing techno-economic modelling as a complementary digital divide regulatory tool – enabling evidence-based interventions – is a strong recommendation too from me.

I introduce and describe techno-economic modelling in Chapter 7 (cf. Section 7.2.5) in the context of BharatNet in India. Earlier in this chapter, I note that the GSMA identified key policy and regulatory blockages in 41 countries in Africa in their 2024 report, namely:

- the lack of monitoring of digital divide and broadband strategies;
- the lack of efficient (i.e., effective and transparent) universal access service funds; and

- the lack of a policy and regulatory framework favourable for the emergence of technological start-ups.

I think techno-economic modelling would certainly be a major help in addressing all these three concerns, too. In this vein, I highlight Dr Christophe Stork's excellent Universal Access Service (UAS) Portal tool whose overview is depicted in Figure 8.3.

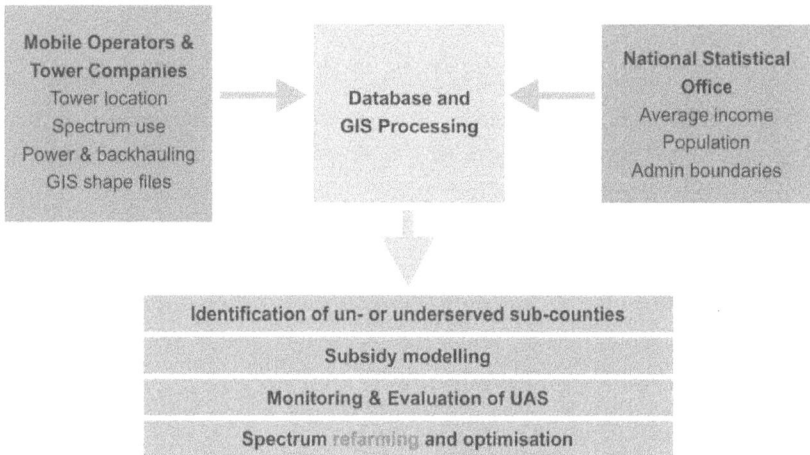

Figure 8.3 – Overview of a Techno-Economic Modelling UAS Portal
Source: Research ICT Solutions Ltd[344], adapted with permission

Dr Stork's Research ICT Solutions have developed regulatory and UAS portals for several clients – and they all have a single point of entry for all sector-specific data, including[345]:

- Fibre and mobile coverage maps from fixed and mobile telcos;
- Population statistics and other administrative data from national statistical organisations and the central bank;
- Operator financials and Key Performance Indicators (KPIs).

---

[344] https://researchictsolutions.com/home/about/our-services/
[345] *https://researchictsolutions.com/home/about/our-services/ - see Regulatory and UAS Portals*

Fixed and mobile licensees input their data using a secure online portal. The inputted data is corroborated using audited financial statements and data from national statistical offices and the central bank. The UAS portals support a wide range of other regulatory tasks, including:

- *Identifying UAS Gaps*: identifying gaps, plan broadband rollout and estimating the costs of extending broadband and voice coverage. This kind of portal contributes to addressing GSMA's finding in 41 countries in Africa of the lack of monitoring of digital divide and broadband strategies;

- *Monitoring and Evaluation (M&E)*: tracking progress toward ICT policy objectives including digital divide goals at very granular administrative levels like State, province, district, village or even smaller. Similarly, this also contributes to addressing GSMA's finding of the lack of monitoring of digital divide and broadband strategies. As Figure 8.3 depicts, such a UAS Portal supports:
  - Identification of un- or under-served counties, sub-counties, districts, or whatever the country governance administrative boundaries are called;
  - Subsidy modelling;
  - Monitoring and evaluation of the UAS funds;
  - Spectrum refarming and optimisation.

- *Predicting ICT tax impact*: estimating the impact of ICT taxes on GDP growth, jobs and net taxation;

- Market share reporting and competition assessment; and

- ICT Sector Performance analyses and reporting.

Such techno-economic modelling portals truly transform how regulators collect and analyse data, allowing them to provide up-to-date and accurate data in real-time to all stakeholders. They are built together with regulatory staff, in order for the latter to train them to maintain and extend the portals independently.

I strongly recommend the employment of such evidenced-based techno-economic modelling tools such as UAS Portals as key to monitoring and directing universal digital divide efforts. They seem to me to address all of the GSMA-identified 'blockages' I enumerate at the beginning of this section.

The ITU also supports such a data-driven or evidence-based regulatory tool along with targeted interventions. ITU-D Sector Director, Dr Cosmas Luckyson Zavazava writes:

> "*Improved data is crucial for bridging the digital divide. Even modest investments in measurement can produce significant returns by pinpointing where needs are most urgent and identifying which interventions will have the greatest impact*[346]. Achieving UMC[347] will require renewed investment in infrastructure, *strengthened policy frameworks, capacity development, and better data.* (*my emphases*)".

I wholeheartedly agree with Dr Cosmas Luckyson Zavazava, and this quote is a very apt summary of some of my key messages in this entire chapter.

## 8.4  Summary

This chapter makes some clear policy and regulatory recommendations that would/may mitigate the connectivity digital divide amongst the 2.6 billion of people *not* connected to the Internet (as of early 2025), and help meaningfully connect 2 billion others.

---

[346] https://www.itu.int/dms_pub/itu-d/opb/ind/d-ind-sddt_afr-2025-pdf-e.pdf, see Foreword
[347] Universal Mobile Connectivity

# 9 Leadership in Bridging the Digital Divide: Insights & Strategies

This chapter covers the sort of leadership required towards mitigating the digital divide, including my recommended unsolicited insights, advice and guidance.

There are several loci of digital divide leadership that need to come together – *synergistically* – for all of them to stand a chance of addressing the 2.6 billion unconnected challenge, and improving on the connectivity of the extra 2 billion *not meaningfully connected*. These loci of leadership include the following non-exhaustive list:

1. Individual country ICT leadership across LDCs, LLDCs, LICS, LMICs and SIDS nations.
2. International Telecommunications Union (ITU) Leadership.
3. ITU Broadband Commission leadership.
4. Telecoms Industry Leadership (including the mobile, satellite, fixed, FWA, etc. industries). Regarding telecoms industry leadership, I must single out the GSMA that represents the interests of mobile network operators and the broader mobile ecosystem, and which unites over 750 operators and almost 400 companies across the globe. The GSMA – like the ITU – publishes key digital divide annual reports too, e.g. their State of Mobile Connectivity report series[348].
5. Multilateral / Intergovernmental (including regional intergovernmental organisations) leadership, e.g. the United Nations (UN), United Nations Development Programme (UNDP)[349], European Commission (EC), Commonwealth Telecommunications Organisation (CTO), APT[350], ATU[351], CARICOM[352], etc.

---

[348] https://www.gsma.com/r/wp-content/uploads/2023/10/The-State-of-Mobile-Internet-Connectivity-Report-2023.pdf
[349] https://www.undp.org/digital/standards/2-bridge-digital-divide - UNDP was the first in the UN System to adopt a "Digital by Default" approach.
[350] Asia-Pacific Telecommunity - https://www.apt.int/
[351] African Telecommunications Union - https://atuuat.africa/about/
[352] Caribbean Community - https://caricom.org/

6. Big Tech (e.g. FANGAM[353]/GAMMA) Leadership.
7. Multilateral development banks (MDBs) Leadership, e.g. World Bank Group which includes the IFC, AfriExIm Bank, etc.
8. Non-Governmental/Private Charitable Organisations (NGOs), e.g. the Internet Society, A4AI[354], Bill & Melinda Gates Foundation, Tony Blair Institute for Global Change, etc.
9. Individual Governmental Aid and Developmental Agencies (USAID[355], UK Aid, Australia Aid, etc.[356]), Aid Agencies & Banks (e.g., China ExIm Bank) – see Section 1.5.

All these above categories of leadership and more are all involved in some way in bridging the connectivity digital divide across the globe. However, addressing the challenge of connecting the unconnected and enhancing meaningful connectivity for billions of people worldwide truly requires a coordinated, multi-stakeholder effort. This is not least because it gets quite confusing – and I have observed some countries 'playing' off one or more category/categories of 'leadership' players against others.

So next, I attempt to present five priority recommendations (maximum) – drawing from all prior chapters – for *some* of these leadership categories involved in mitigating the digital divide. I do this to be both succinct and direct. By 'direct', I mean that they are arguably not the 'traditional' recommendations the reader may expect – I have covered those in previous chapters. Rather, these are frank and perhaps 'undiplomatic' in cases.

This is the chapter that gives this book one of its sub-titles on the inside cover of this book: "*Unsolicited Insights, Advice and Guidance to Developing Countries, the Broadband Commission, the ITU, the World Bank Group, UN Agencies and more*". The more general form of this sub-title at the front cover of the book reads: "*Expert insights for global institutions and developing nations committed to achieving universal digital*".

---

[353] Facebook (Meta), Amazon, Netflix, Google, Apple and Microsoft
[354] Alliance for Affordable Internet – www.a4ai.org
[355] U.S. Agency for International Development (USAID) – as of March 2025, this eminent organisation is being 'dismantled' by the 2nd-term Donald J. Trump Administration.
[356] See Section 1.5, Chapter 2

## 9.1 Individual Intra-Country Leadership

Generally, I find true digital divide, universal connectivity leadership across LDCs, LLDCs, LICS, LMICs and SIDS nations very lacking. I acknowledge that this generalisation may read as unevidenced. However, I believe the data from reputable organisations support my thesis:

- Most of these nations have broadly relegated their digital divide challenges to their 'telecoms markets' to 'own' and address, literally and lazily hoping that all 'would be fine on the night'. Even when the ICT leaders of these countries are informed of the evidential scale of the coverage and usage digital divide gaps for their countries, they appear largely oblivious to them, and to the complexity to address them. How else do we explain the fact that telecoms market has failed key [rural] populations – with a huge global usage gap of circa 1.3 billion [minimum] *not* regularly using basic 2G voice services as of March 2025 [cf. Section 6.1.1]?

- I regularly see 'broadband plans' in many LDCs that are hardly 'living' documents, i.e. they are *not* guiding and directing any digital divide mitigating efforts inside many of the countries. Whilst many of these LDCs have broadband plans, the ITU correctly observes that their implementation [as plans] are often perfunctory – they are not always effectively executed due to factors such as limited funding, lack of technical expertise, and insufficient efforts to realise them[357]. The Broadband Commission also confirms this thesis[358].

- I regularly observe Universal Service Funds (USFs) that are very poorly managed and who hardly deliver concrete outcomes to mitigate the digital divide in these countries – yet they are hardly reformed. Research by A4AI in 2018 showed that a significant portion of USF funds often remains unspent. In Africa, A4AI reported that over $400 million collected for Internet access expansion was left unused, which could have connected millions of people[359]. Not much has changed from my observations of USOFs since 2018.

---

[357] https://www.itu.int/en/ITU-D/LDCs/Pages/ICTs-for-SDGs-in-LDCs-Report.aspx
[358] https://www.broadbandcommission.org/ldc5-dpoa/
[359] https://a4ai.org/research/report/universal-service-and-access-funds-an-untapped-resource-to-close-the-gender-digital-divide/

So, here are my five concise primary recommendations for addressing the digital divide, specifically directed at the *leadership* of individual least developed countries (LDCs):

1. *Take the problem seriously & authentic leadership*: I do not think most of these countries do take digital divide leadership seriously at all like in India. Individual country ICT leadership across LDCs, LLDCs, LICS, LMICs and SIDS nations <u>must mean what they say</u> when they say they want to bridge the digital divide in their countries. Otherwise, they are wasting so many intra-country stakeholders' time. By failing to prepare, you are preparing to fail. LDC countries must truly develop deeper strategies to bridge the digital divide usage and coverage gaps. Relying solely on market-driven solutions overlooks systemic challenges such as affordability, digital literacy, infrastructure development in rural areas, and inclusive access.

   True digital divide leadership is no different from any other good true leadership. Leaders shape goals and develop novel ideas. Leaders connect with others at an emotional level to ensure the vision, goals, and novel ideas are truly shared and owned. It is such TMT/ICT leadership that has delivered some of the most TMT/ICT-successful economies in the world, including South Korea, Japan, and Singapore.

2. *Prepare for a long haul, hard graft to eliminate coverage gaps, build fibre backbones and more*: these countries must learn from the Pareto principle that predicts or informs us that most of the real digital-inclusion 'hard graft' efforts are mostly still ahead of us (cf. Section 1.3) – some 86% of the efforts needed is still ahead of us. So like with the Indian rural connectivity case study (see Section 7.2), they really must dig in for the long haul to address their digital divide challenges. Governments and ICT leaders need to play an active role by collaborating with both private and public sectors, investing in long-term policies, and focusing on widespread accessibility. The sustained implementation of a long-term policy along with creative and equally sustained *regulating* are key (see Chapter 8). *Independent* regulators may be best placed to regulate to mitigate the digital voice and Internet/data divide over much longer periods – beyond the typical 4-to-5-year electoral cycles in most countries. It is a truly <u>long-haul graft</u> building out from the *first*-mile [typically] submarine cables – where the Internet enters a country – building out the *second*-mile country backbone, and then the *last*-mile

reaching to the end users and businesses – and all the *invisible*-mile numerous components of the Internet. Do not worry, I describe all these 'miles' later in Section 9.4.

3. *Prioritise the usage gap challenge intra country*: the usage gap challenge is exponentially harder that the coverage gap one, yet many countries just ignore it in preference for the easier coverage gap challenge. Usage gaps contributors like crushing poverty, general poverty (low income), lack of awareness of the Internet/data, connectivity being a low-priority concern, illiteracy, lack of digital skills, low population density (rural areas), usage-based pricing models, poverty, miscellaneous issues, etc. are truly non-trivial. Broadband plans *hardly* address usage gap issues adequately because they are too hard (see Section 2.5). Countries need to redress this.

4. *Consider 'Civilisation-in-a-box'-type solutions towards addressing the coverage gaps*: solutions of this ilk attempt to 'kill several birds with one stone' (cf. Chapter 4).

5. *Collaborate both intra-country and inter-country*: *If you want to go fast, go alone. If you want to go far, go together*. The wisdom of this African proverb is profound. The proverb speaks to the power of collaboration and unity which I cover this in much detail in Chapter 7.

Ultimately, there is virtually no substitute for determined and concerted in-country leadership, and stopping all the platitudes from ICT Ministers, Regulators and leaders of developing nations that fool only themselves. I read a recent headline: *Rwanda intends to connect every home by 2029*[360]. The Rwandan ICT Minister Ms. Paula Ingabire announced in March 2025 that:

"This is one of the primary goals of the government's second national transformation strategy (NST2 2024-2029), which aims to achieve the Vision 2050 blueprint of sustainable economic growth, prosperity, and a high standard of living for all citizens… We want to get to a point where, when it comes to last mile connectivity, *every home in Rwanda at least*

---

[360] https://itweb.africa/content/KA3WwqdzKb57rydZ

*is well connected and they are able to benefit from the digital economy.*"[361] (my emphases).

I like this ambition, and being Rwanda, they may be able to pull it off. However, Rwanda should be realistic: recall and the reader can see from Table 6.3 (Chapter 6) that the national meaningful connectivity in 2021/2022 was only 0.6% in Rwanda compared to 26.2% in Colombia (A4AI, 2022). Recall too that an IFC/World Bank 2024 survey of 3,325 microenterprises in seven African countries found the use of the Internet for business purposes was 7% on average, ranging from 24% in South Africa to 1% in Rwanda (see Section 8.2.2). Etc.

Rwanda is a landlocked nation making 'first mile' issues challenging, but they are small geographically to make the 'middle mile' challenge more doable. The 'last mile' challenge will be much harder. However, ultimately, they need to learn the lesson that *usage gap* issues are much more intractable than the coverage gap issues that this ambition is striving to address. Nonetheless, I emphasise that such concerted in-country leadership is an absolute must to bridge the digital divide. Rwanda should learn from the insights and recommendations in this book.

## 9.2  ITU Leadership

The ITU plays a pivotal role by fostering global collaboration and implementing initiatives towards bridging the digital divide. Recall from Chapter 1 that the ITU has three main sectors: ITU-R (Radiocommunication), ITU-T (Standardisation), and ITU-D (Development). The ITU uniquely leads on the following:

- *Research and Data*: the ITU's research yielding their annual Facts and Figures Report on the digital divide – and their interpretations – are both invaluable and almost irreplaceable, e.g. the ITU's 2024 Facts and Figures 2024 report[362]. The reader may recall from the beginning of Chapter 1 where I cite the ITU Secretary-General (SG), Doreen Bogdan-Martin commenting

---

[361] *Ibid.*
[362] https://www.itu.int/itu-d/reports/statistics/2024/11/10/ff24-mobile-phone-ownership/

on the ITU's Facts and Figures 2024 report[363] as "a tale of two digital realities between high-income and low-income countries".

- *Advocacy for Digital Inclusion*: as can be gleaned and projected from SG Bogdan-Martin's quotation above, the ITU advocates for the inclusion of marginalised groups into the digital world, as well as for gender equality, e.g. the ITU's support of the EQUALS in Tech Awards that highlight gender digital divide[364].

- *Global Policy Development*: the ITU develops global policies and strategies towards expanding ICT access by working with member states, particularly targeting rural underserved regions. The policies include promoting universal voice and broadband access and affordable connectivity.

- *Training & Capacity Building*: the ITU provides much training, capacity building and relevant resources to build digital skills in LDCs, to empower SMEs and citizens to employs and use ICTs effectively. This is especially invaluable to Least LDCs where digital literacy is often exceptionally low.

- *Digital Infrastructure Support*: the ITU facilitates the deployment of ICT infrastructure through offering technical expertise and fostering public-private partnerships, to bridge the gap in regions with limited connectivity.

- *Global Radio Spectrum Regulator and Global Standards Body*: through its Radiocommunication Sector (ITU-R), the ITU oversees the allocation and management of the radio-frequency spectrum and satellite orbits, ensuring that various services including mobile, broadcasting and satellite communications coexist without interference[365]. The ITU develops and maintains the Radio [Spectrum] Regulations, the key international treaty that governs the *equitable* use of the radio spectrum and satellite orbits. The ITU Standards Sector (ITU-T) develops technical standards to ensure seamless interconnection of networks and technologies worldwide[366]. These standards cover a wide range of areas, including terrestrial and satellite-based communication systems.

- *Raising Digital Connectivity Funds*: the ITU has recently called for and is raising USD 100 billion in commitments to its Partner2Connect (P2C)

---

[363] *Ibid.*
[364] https://www.equalsintech.org/
[365] https://www.itu.int/en/mediacentre/backgrounders/Pages/itu-r-managing-the-radio-frequency-spectrum-for-the-world.aspx
[366] https://www.itu.int/hub/2020/10/developing-global-standards-for-radio-based-telecommunications/

Digital Coalition. The P2C Digital Coalition is ITU's global alliance to mobilise and announce new resources, partnerships, and commitments for universal, meaningful connectivity. P2C drives digital transformation across all countries, with special attention to the hardest-to-connect communities, particularly in Least Developed Countries (LDCs), Landlocked Developing Countries (LLDCs), and Small Island Developing States (SIDS)[367].

In one of its key aims to create a more inclusive global digital society where everyone – regardless of location or socioeconomic status – can benefit from the opportunities offered ICT and mobile/satellite communications, the ITU is truly irreplaceable.

Nevertheless, despite the ITU's irreplaceable role in the digital divide, here are my five top recommendations to them on mitigating the digital divide:

1. *Continue to do all the above – but the ITU should now truly learn from both the Pareto principle and from the realities of the 2.6 billion unconnected*: I applaud the ITU for doing what they do towards bridging the digital divide world-wide as I overview above. However, I strongly recommend they too learn from the Pareto principle that predicts or informs us that most of the real digital-inclusion 'hard graft' efforts are mostly still ahead of us (cf. Section 1.3) – some 86% of the efforts needed are still ahead of us. I recommend their global policies are more informed by the realities of the current 2.6 billion unconnected in countries like PNG and Malawi (e.g. see Sections 2.2 and 2.3).

2. *Work with Member States to set more realistic policies and targets*: drawing from the latter – and as I note several times already earlier in this book – there are stretch digital inclusion targets vs. patently impossible ones (cf. Section 2.1.1). The ITU should work with LDCs, LICs, LIMCs, SIDS, LLDCs and even EDCs to set more developmental category-led and measurable, realistic targets and policies.

---

[367] https://www.itu.int/partner2connect/ - Partner2Connect

3. *Revise their [ITU's] definition of meaningful connectivity and adopt Research ICT Africa's*[368] *After Access Survey methodology*: I recommend the ITU constructs a more appropriate 'meaningful connectivity' question that provides a more nuanced and comprehensive picture by incorporating different questions in their survey than they have today – as I propose in Section 1.2.

This is because I believe that the meaningfully-connected 'mid tail' of Internet users in 2025 (refer to Figure 1.1) – according to the definition established by the ITU – is likely to constitute only a fraction of the 2.7 billion users, potentially even less than 40%. The ITU should prove me wrong. I assert in this book that circa 2 billion of the so-called connected are *not* meaningfully connected at all. My thesis is bolstered and supported by the reputable A4AI's researched estimates of meaningful connectivity, by country, geography, and gender in several countries (see Table 6.3 in Chapter 6) that the national meaningful connectivity in 2021/2022 ranging from 0.6% in Rwanda to 26.2% in Colombia (A4AI, 2022).

In this vein, I encourage the ITU and readers to consult and consider the Research ICT Africa (University of Cape Town) *Alternative Policy and Regulatory Strategies for Developing Countries in the Digital Economy* policy brief[369]. The Research ICT Africa centre (led by Dr Alison Gillwald) has precisely been challenging this ITU data and definitions of *meaningful access and use*. Their works provide alternative national representative data on Internet access and use, using their well-thought out *After Access Survey*[370] that demonstrates unequivocally that the main barriers to access and use are demand-side usage gap challenges, not supply-side issues. Their survey has been instrumental in showing that digital inequality is not just about infrastructure but also about affordability, digital literacy, and socio-economic factors[371]. Yet the ITU and everyone else follows the ITU lead

---

[368] https://afteraccess.net/
[369] https://researchictafrica.net/2018/07/13/alternative-policy-and-regulatory-strategies-for-developing-countries-in-the-digital-economy/ - Alternative policy and regulatory strategies for developing countries in the digital economy, Research ICT Africa
[370] https://researchictafrica.net/project/after-access-2022-survey/
[371] https://afrisig.org/sites/default/files/pdf/Acess-After-Access-AfriSIG-2024.pdf

and continue to spend millions of dollars on supply-side digital divide solutions.

Research ICT Africa – similar to some of my recommendations in this book – "calls for alternative policy, regulatory, business, community, cooperative, and human development strategies that respond *to the limited but definitive evidence that education and income are the primary determinants of Internet access and use*"[372] (*my emphases*). They have long contended that we cannot continue to do the same thing and expect different results. We cannot ignore the demand-side evidence because we only have supply-side solutions.

4.  *Global Digital Divide Strategy, Collaboration, Policy, Regulation & Leadership (cf. Chapters 7 and 8)*: I can think of no better and more competent and global multilateral organisation to lead on my proposals in Chapters 8 and 9 than the ITU.

5.  *ITU-D would do better if they adopt and own ALL the insights in this humble book*: the International Telecommunication Union's Development Sector – ITU-D – can do better. This third sector – after ITU-R and ITU-T – was established in 1992 with the primary goals to promote digital development and bridge the digital divide by leveraging information and communication technologies (ICTs)[373]. Whilst ITU-D has made significant strides in promoting digital development and bridging the digital divide, particularly in underserved regions[374], I hope this book 'proves' that too many challenges persist, such as the stark gaps resulting in over 90% of the 2.6 billion people offline residing in developing countries.

The reality is that the scale of global inequality in digital access suggests to me that much deeper solutions are required for both supply-side gaps, and even more so for demand-side challenges. Current ITU-D Director, Dr Cosmas Luckyson Zavazava, observed of the ITU's own Facts and Figure 2024 that:

---

[372] Personal communications with Dr Alison Gillwald of the Research ICT Africa (University of Cape Town), as part of her peer review of this manuscript/book.
[373] https://www.itu.int/en/mediacentre/backgrounders/Pages/itu-d-driving-ict-led-development-worldwide.aspx
[374] https://www.itu.int/en/ITU-D/Pages/ITU-D25/achievements.aspx

"The world *is inching towards universal access at a time*. While we continue to make progress on connectivity, our advances *mask significant gaps in the world's most vulnerable communities*, where digital exclusion makes life even more challenging. We must intensify our efforts to remove the barriers that keep people offline and close the usage gap, and renew our *commitment to achieving universal and meaningful connectivity*, so that everyone can access the Internet[375]" [*my emphases*].

I agree we are only "inching" indeed, though Pareto's prediction [see Section 1.3] is much more pessimistic. With the benefits of the hindsight, since 1992, ITU-D is now fully aware of the true depth of the *usage and coverage gaps* challenges, as well as the funding limitations, geopolitical complexities, and rapid technological advancements – that unfortunately are widening the digital divide instead of narrowing it. I am glad to see the ITU-D Director mention 'meaningful connectivity' too. Indeed, in an April 2025 ITU publication, Dr Cosmas Luckyson Zavazava declares that "universal and meaningful connectivity (UMC) is a new imperative"[376]. So, I recommend the ITU-D considers, 'owns' and adopts all the insights and suggestions in this book.

## 9.3   ITU Broadband Commission Leadership

The distinction I am making here between the ITU and the ITU's Broadband Commission *may* be one without a difference in terms of my recommendations. However, the Broadband Commission[377] was launched in 2010 by the ITU and

---

[375] https://www.itu.int/en/mediacentre/Pages/PR-2024-11-27-facts-and-figures.aspx
[376] https://www.itu.int/dms_pub/itu-d/opb/ind/d-ind-sddt_afr-2025-pdf-e.pdf, see Foreword
[377] https://www.broadbandcommission.org/publication/state-of-broadband-2023/ - "The Broadband Commission for Sustainable Development was originally launched in May 2010 by the International Telecommunication Union (ITU) and the United Nations Educational, Scientific and Cultural Organisation (UNESCO) as the Broadband Commission for Digital Development. It is comprised of top industry leaders, government leaders, international agencies and development organisations. Commission members work together to devise strategies that advocate for higher priority to be given to the

UNESCO to deliberately include top industry leaders, Government leaders, international agencies and development organisations. The Broadband Commission has striven to put universal broadband connectivity [for all humans on earth] at the forefront of global policy discussions through developing practical and sustainable policy recommendations in order to accelerate progress towards achieving the United Nations (UN) 2030 Agenda, and the UN's own 17 Advocacy Targets[378].

So, there may be some difference in the sense that the Broadband Commission – with its top industry leaders, government leaders like President Paul Kagame of Rwanda, academia and the involvement of other international agencies like UNESCO – is a unique *public-private partnership (PPP)* that develop actionable recommendations and targets that promote broadband as a key driver to Sustainable Development. The ITU has a much broader mandate with its three main sectors: ITU-R (Radiocommunication), ITU-T (Standardisation), and ITU-D (Development).

So, my five top recommendations to the PPP Broadband Commission that primarily promotes broadband are the following:

i. *Understand, internalise and advocate for the recommendations I make above for the ITU*: the Broadband Commission is clearly a 'daughter' or 'son' of the ITU working in the broader connectivity 'family business'. So, the five recommendations I make above for the ITU arguably and rightfully apply to the Broadband Commission too.

ii. *Review the Broadband Commission's targets-setting process:* I argue in Chapter 2 that the 2018 Broadband Commission Targets for 2025 (cf. Section 2.1.2) were implausible for most LDCs. I note that perhaps LDCs, LICs, LIMCs, SIDS, LLDCs and even EDCs should be the key nations to influence and set Broadband Commission target dates.

iii. *Help Facilitate truly innovative 'lever' and 'pulley'-type coverage gaps solutions*: it appears to me that with the Broadband Commission – being

---

development of broadband infrastructure and services, to ensure that the benefits of these technologies are realised in all countries, by all people" – Source: www.BroadbandCommission.org

[378] https://www.sightsavers.org/policy-and-advocacy/global-goals/

a true PPP (that includes the involvements of academia, top industry and Governments) – can help facilitate and productise 'innovations' like Civilisation-in-a-box (cf. Chapter 4). The Commission may be able to set up competitions and grant programmes that promotes more innovative solutions to the true realities that the civilisation-in-a-box 'solution' strives to address. Then the Commission may help facilitate other companies to productise the innovative solutions. I honestly believe that if the coverage gap broadband questions are well framed, or better-framed than I do in Chapter 4, innovative solutions and products would naturally emerge.

iv.  *Help seek truly deeper policies, solutions and partnerships to the much harder usage gap challenges*: the usage gap (demand-side) challenges are so much harder to address. For example, as I was penning this chapter, I saw a news story with the headline: *710m people in sub-Saharan Africa can't afford handsets*[379]. This presents a true usage gap conundrum. According to the story, the Africa Group of Six[380] (G6) says that while they have made considerable progress in their collective goal to expand mobile coverage in sub-Saharan Africa, 60% of people who live in covered areas *cannot afford* handsets to use mobile services. This G6 met with key stakeholders from the telecoms industry, global institutions and Governments at the Mobile World Congress 2025 in Barcelona, with much of the focus of the discussion on handset affordability. The G6 asserts that this handset affordability 'culprit' is largely responsible for the 60% usage gap, which represents 710 million people who live within network coverage areas but are not subscribing to mobile Internet services. Even low-end smartphones remain too expensive for them. This is a problematic issue that seems to me well suited to the Broadband Commission's facilitation.

The G6 operators also want to see the lowering of taxes on low-end smartphones to cover 52 African nations, *inter alia*, by convincing the United Nations Economic Commission for Africa (UNECA) to get

---

[379] https://developingtelecoms.com/telecom-business/telecom-regulation/18103-710m-people-in-sub-saharan-africa-can-t-afford-handsets-g6.html
[380] A coalition of African mobile network operators (Airtel, Axian Telecom, Ethio Telecom, MTN, Orange, and Vodacom)

involved. The G6 claims to have engaged the World Bank Group, the African Development Bank (AfDB), the ITU, the GSMA Handset Affordability Coalition and more. They are also advocating that the World Bank Group expands electricity access to 300 million Africans by 2030, concentrating on green energy solutions, particularly off-grid ones, and underserved areas. Recall that we are still faced with a 1.294 billion usage gap for basic [2G] voice (cf. Table 6.1).

The Broadband Commission could significantly impact the following areas: electricity, handset affordability, digital literacy, lack of awareness, connectivity as a low-priority concern, lack of digital skills, usage-based pricing models, and other miscellaneous issues.

v.    *Other miscellaneous goals to be championed by the Broadband Commission*: perhaps, since its inception in 2010, the Commission should review its mandate. I believe there are broadband-related areas that the Broadband Commission appears eminently suitable to champion:

- Broadband Public-Private Partnerships: toward ensuring affordable broadband Internet access.
- Policy and Regulatory Frameworks: such as the tax recommendation on affordable handsets promoted by the African G6.
- International Cooperation: the Commission could best facilitate multilateral organisations like the United Nations, World Bank, the ITU, UNECA, GSMA, top tech companies (like Meta, Microsoft, Google, etc.), top Satellite companies (like Starlink, Inmarsat, etc.) – fostering true international cooperation, sharing best practices, and providing financial and technical support.
- Facilitate NGO Involvement: Non-governmental organisations can work with the Commission to raise awareness and provide digital literacy training to SMEs and communities.
- Innovation and Technology: the Commission can help facilitate innovative broadband technologies and solutions such as civilisation-in-a-box, satellite Internet, wireless FWA networks, promote hundreds of ISPs like in Brazil and community-based

networks. These would all help reach areas where traditional infrastructure is not feasible.

Whatever the new Broadband Commission mandates are – it feels to me time to review and add to them.

## 9.4   Telecoms Industry Leadership

I am sure I would (or have) come across in this book thus far as a major critic of the mobile telecoms industry, for having failed to connect the 2.6 billion people to date as of early 2025. The reality is different: having lived through the near-complete failure of the Fixed Telecoms industry to connect most of the globe prior to the beginning of 1990s, I am so much enthralled about what the Mobile/Cellular industry has achieved over the last 25 to 30 years. I am in print already (see Nwana, 2022, Section 2.3.6.1) stating boldly that *The Mobile/Cellular Industry has literally changed the world*. I mean it. As I note in my 2022 book (Nwana, 2022) – as someone who hails from Central/West Africa [Cameroon] – the mobile industry has just been a revolution to the lives of people there. For example, the ICT sector of one of the largest economies in Africa (Nigeria) is dominated by the telecoms industry which contributed 14.4% to total Nominal GDP during Q4 2024, and the ICT sector contribution to nominal GDP in Q4 2024 in Nigeria was recorded as 17%[381]. For comparison, telecommunications contributed a measly 0.06% to national GDP in Nigeria in 1999, just over 25 years ago, and was only 3.5% fourteen years ago in 2011 (Nwana, 2014, p. 10). No Nigerian or sub-Saharan African would or should quibble about the positive externalities that have accrued from the mobile sector in Nigeria and Sub-Sahara Africa. According to the GSMA, the mobile/cellular industry as of 2024 had connected 5.8 billion subscribers *uniquely*[382] and 4.7 billion Internet subscribers. This has been no mean feat. So, I am truly and personally thankful to the *mobile* telecoms sector contributions in Africa over the last 25 years.

---

[381] nigerianstat.gov.ng & https://nairametrics.com/2025/02/26/ict-contributes-17-to-nigerias-gdp-in-q4-2024-despite-slower-growth-nbs/
[382] https://www.gsma.com/solutions-and-impact/connectivity-for-good/mobile-economy/wp-content/uploads/2025/02/030325-The-Mobile-Economy-2025.pdf

In the same vein, I was critical in the same book (as in this one) about the overpromises and lack of delivery of the satellite telecoms industry. In 2023, the global space/satellite economy was valued by McKinsey at approximately USD $460 billion[383]. In contrast, the mobile sector contributed in 2023 around 5.8% of global GDP, amounting to USD $5.7 trillion according to the GSMA[384], i.e. 12+ (twelve+) times bigger.

As I point out clearly in Section 6.3 [Chapter 6], for the many reasons I provide in that section, the reality is that satellite will/would continue to remain very *niche* to connecting the 2.6 billion not online without significant [Government] subsidies.

So, most of the telecoms industry leadership to address the coverage and usage gaps today [and the 2.6 billion unconnected as well as the 2 billion not meaningfully connected] would be needed from the mobile/cellular sector, despite the continued promises from the satellite industry. If anything, Wi-Fi - a local wireless networking technology that connects devices within a limited range (like a home or office) to a router or access point – is arguably much more important. The Wi-Fi Alliance® commissioned and published a seminal report that estimated the annual global economic contribution of Wi-Fi® at US$ 1.96 trillion, projecting that number to surpass USD 3.47 trillion by 2023[385].

I also think that role of fixed telecoms is largely underestimated, understated and becoming increasingly critical. The reason is simple, and is illustrated in Figure 9.1 from the World Bank which describes the concepts of first-, middle-, last- and invisible miles. I alluded to or mentioned these 'miles' earlier in Section 9.1.

The 'first mile' refers to where the Internet enters a country, i.e. the international Internet access including submarine cables, landing stations, satellite gateways and dishes, cross border microwave and other fixed fibre type links across the borders. The 'middle mile' – as Figure 9.1 denotes – refers to where the Internet passes through the country in question, notably the national [typically fibre]

---

[383] https://www.mckinsey.com/industries/aerospace-and-defense/our-insights/space-the-1-point-8-trillion-dollar-opportunity-for-global-economic-growth

[384] https://www.gsma.com/solutions-and-impact/connectivity-for-good/mobile-economy/wp-content/uploads/2024/02/260224-The-Mobile-Economy-2024.pdf

[385] https://www.wi-fi.org/news-events/newsroom/wi-fi-global-economic-value-reaches-196-trillion-in-2018

backbone, inter-city fibre networks, metro intra-city networks, Internet eXchange Points (IXPs), local hosting of content and more. The 'last mile' is where the Internet reaches the end-user, including Local Access Networks (LANs)/Wi-Fi, local loops, telecoms exchanges, wireless (3G/4G/5G) masts, etc. Then there is the 'invisible mile' which refers to all the hidden elements that are equally vital to the integrity of the entire Internet value chain – this includes all our invisible radio spectrum, network databases, cybersecurity processes, *policy and regulatory reforms* and more. These telecoms industry-provided 'miles' are truly critical to what we now know as the world's Internet infrastructure.

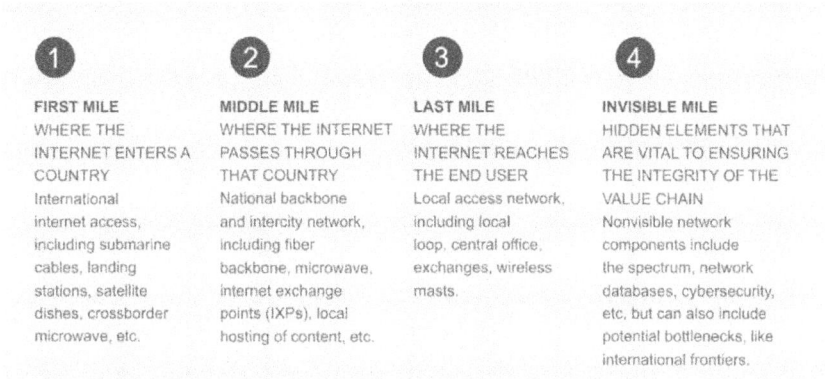

| ① | ② | ③ | ④ |
|---|---|---|---|
| **FIRST MILE** | **MIDDLE MILE** | **LAST MILE** | **INVISIBLE MILE** |
| WHERE THE INTERNET ENTERS A COUNTRY | WHERE THE INTERNET PASSES THROUGH THAT COUNTRY | WHERE THE INTERNET REACHES THE END USER | HIDDEN ELEMENTS THAT ARE VITAL TO ENSURING THE INTEGRITY OF THE VALUE CHAIN |
| International internet access, including submarine cables, landing stations, satellite dishes, crossborder microwave, etc. | National backbone and intercity network, including fiber backbone, microwave, internet exchange points (IXPs), local hosting of content, etc. | Local access network, including local loop, central office, exchanges, wireless masts. | Nonvisible network components include the spectrum, network databases, cybersecurity, etc, but can also include potential bottlenecks, like international frontiers. |

Figure 9.1 – Defining First, Middle, Last & Invisible Miles
Source (World Bank[386])

Let me revert to why I bothered to introduce and describe all these 'miles'. I did so largely to draw attention to the fact that the first and second mile of telecom networks – i.e., connecting countries, regions/states/provinces in countries, homes, businesses, and infrastructure to the broader Internet network - remain heavily reliant on fixed-line technologies, such as fibre optics and copper cables. This "fixed" foundation is vital for ensuring high-speed, reliable broadband connectivity. For example, Fibre-To-The-Home (FTTH) deployments, are

---

[386] "Innovative Business Models for Expanding Fiber-Optic Networks and Closing the Access Gaps (Vol. 1 of 3): Main Report." Available at https://documents.worldbank.org/en/publication/documents-reports/documentdetail/674601544534500678/innovative-business-models-for-expanding-fiber-optic-networks-and-closing-the-access-gaps

increasingly prioritised due to their ability to handle the growing demands of modern data usage. Fixed networks have made and are making a huge return – the submarine cables connecting the world's continents are all 'fixed'.

### 9.4.1 Mobile Telecoms: The Pareto Limits and 5G's Role in Connectivity

Drawing from the narrative in Section 9.4 thus far, here are my three top recommendations to the telecoms industry leadership on mitigating the digital divide, which spreads across Sections 9.4.1 and 9.4.2:

i. *The Mobile Telecoms Industry – who are critical to the last mile - should be more honest with Developing Countries' Leaderships on the [Pareto] limits of their Business models:* the mobile Industry is by far the greatest contributor and most important stakeholder in most LDCs as regards the last mile connectivity, i.e. how the Internet (even 2G voice) reaches the end customer. However, the mobile industry has reached the [Pareto] limits of its profitable business model in many/most developing markets.

What does this mean I can hear you the reader ask? Well, I read (in March 2025) an interview that Bharti Airtel CEO, Gopal Vittal, gave in which he insisted that the mobile/cellular operator's strategy of allocating capital to fewer areas to improve service quality has been key to its financial turnaround. He added that *Airtel had made a "very brave call" in deciding to go after 40 per cent of the customers which account for almost 80 per cent of revenue*[387]. This honest CEO has said the quiet part out loud.

What Bharti CEO Gopal Vittal said out loud is a variation of the economics of the mobile/cellular industry, which is depicted in Figure 9.2.

I am referring to the classic economics of the mobile telecoms industry where a small proportion of sites – roughly 10-20% – are responsible for generating half (40-50%) of the total revenue. This type of distribution,

---

[387] https://www.mobileworldlive.com/airtel/airtel-ceo-underscores-strategy-shift-for-turnaround/ OR https://www.mobileworldlive.com/airtel/airtel-ceo-underscores-strategy-shift-for-turnaround/ AND https://www.mobileworldlive.com/old_latest-stories/interview-bharti-airtel-group/

frequently encountered in business contexts, corresponds with the Pareto Principle (80/20 rule), as illustrated in Figure 9.3, where a small percentage of contributors are responsible for a substantial portion of the outcomes.

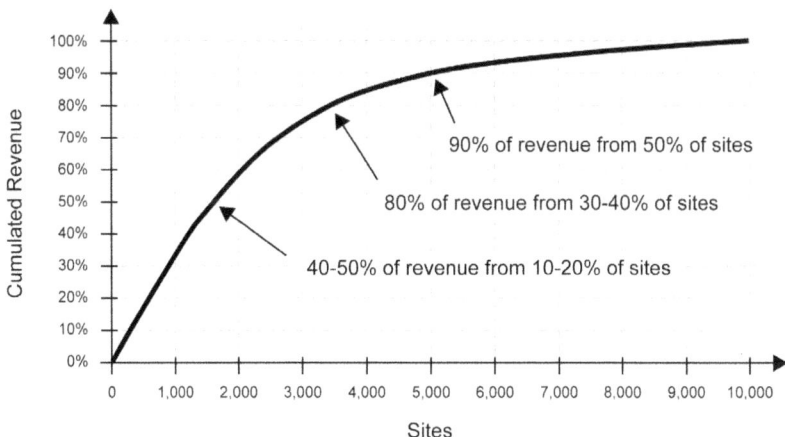

Figure 9.2 – Large share of revenue comes from few mobile sites
Source: Author from miscellaneous reports

Vittal confessed *that 40 per cent of Airtel's subscribers account for almost 80 per cent of Airtel's revenues.* Guess where the 40% of subscribers live? Yes, you guessed right – in the urban areas. In PNG and Malawi (cf. Sections 2.3 and 2.4] where only 15% of their populations live in urban areas, the reader can predict that the mobile industry there would mainly concentrate on the urban populations. If, as Figure 9.2 shows, 80% of the revenues emanate from only 30-40% of the sites – which would largely reside in the urban areas of developing countries – this is clear evidence of the profitability limits of mobile operators. Bharti's Vittal says the Airtel strategy is to go after the 40% of subscribers who generate 80% of the revenues. *What does the reader think this means for the other 60% of subscribers – or more accurately in developing counties – 'to be subscribers'?*

Interestingly, I draw the attention of you, the reader, to Figure 9.3 – the Pareto Distribution Curves.

If you study the curves carefully – as shown in Figure 9.3 - you will note that the '40% of subscribers generating 80%' of revenues' maps to, or follows closely, the 70/30 Pareto curve. This curve also predicts that 30% of the subscribers generate 70% of the revenues. The curve further predicts that 60% of Vittal's Airtel subscribers would broadly generate 90% of their revenues. I think this is a solid prediction. More bluntly, there is another solid prediction that the last 40% of the subscribers (from 60%-100%) only generate 10% of the revenues. The reader should refer to Figure 9.2 wherein the last 10% of revenues come from 50% of the sites. Mobile/cellular companies know these numbers, and they dictate what rational profit-maximising mobile/cellular operators would do – indeed what they do. These are what I refer to as the Pareto Limits of the mobile industry.

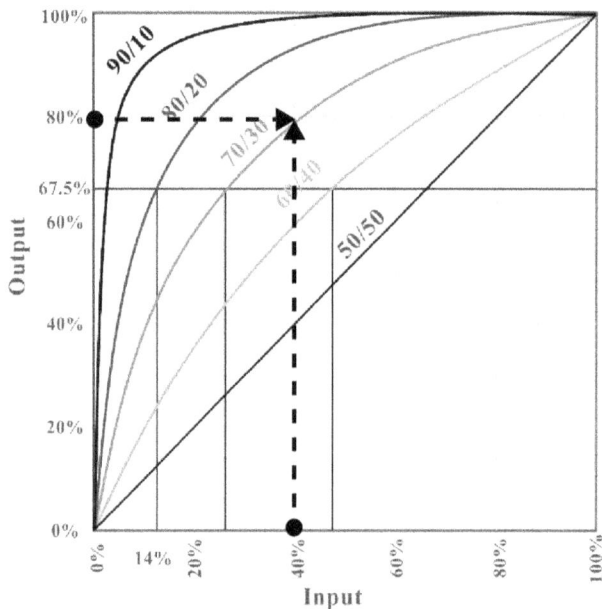

Figure 9.3 – '40% of subscribers generating 80%'
maps closely to the 70/30 Pareto Curve

Referring to Gopal Vittal's home telecoms market [India], Vittal is blunt in describing the environment before consolidation as "a competitive onslaught", noting that industry revenue had collapsed some 30%, with ten players exiting the "market under various circumstances"[388]. India is still

---

[388] *Ibid.*

largely a developing country, albeit an Emerging and Developing Country (EDC). If India's telecoms firms are facing such Pareto limits, then so would most telecoms firms in sub-Saharan Africa, Southeast Asia and the poorer nations in Latin America and the Caribbean.

So, I would like to see the leadership of mobile/cellular companies be honest with ICT Ministers and Regulators of developing countries about what Vittal of Bharti Airtel said in the interview, and for them to publish their equivalent Figure 9.2 and Figure 9.3 numbers for their countries. These would yield Pareto limit curves – evidence that would help inform policy makers and regulators of developing economies of the true magnitude of their challenge, and therefore how much regulatory intervention is necessary.

Whilst at these, the mobile/cellular industry could also be more honest in telling developing countries that 5G (and 6G) is *not* the answer to connecting their proportion of the 2.6 billion unconnected. I acknowledge that mobile operators are likely to play a significant role in enhancing meaningful connectivity for the approximately 2 billion people who are projected to remain without meaningful connections by 2025.

### 9.4.2 Shared Fixed Networks: Partnerships for First- & Middle-Mile

ii. *Big [Fixed & Mobile] Telecoms national companies should continue to invest in and maintain the world's first mile infrastructure. However, they should also build [or be allowed to build] the extensive middle mile backbones too, in order to help monetise their first mile investments. Alternatively, these fixed and mobile telcos should work in partnership with Governments, Universal Service Funds (USFs) and other private investors to build the middle mile backbone networks*: traditional telecommunications providers such as British Telecom (BT), France Telecom (FT), AT&T, Verizon, Deutsche Telekom, MTN, Vodacom, Telkom (South Africa) and China Telecom own significant portions of subsea/submarine cables.

These 'big telcos' often collaborate in international consortiums to build and maintain these networks. It is also true that many submarine first mile cables are owned by consortia of stakeholders, including big telecom operators,

Governments and private companies, who share the costs and responsibilities[389].

Every new submarine cable than lands in a LDC country (typically) leads to an increase in the international Internet bandwidth used to deliver services to populations. The Governments of developing economies truly should all know about this following important evidence: a relatively recent World Bank study published in July 2024[390] demonstrates that each doubling of such 'first mile' capacity in Africa generally results in a 7% drop in the price of fixed broadband internet in Africa. It is an even greater drop for the price of mobile broadband, at circa 13% across the African continent. Using Africa again to illustrate, the ITU found that that the costs of fixed internet and mobile data in Africa represented 14.8% and 4.48% of GDP, respectively, in 2023[391]. These figures are well above the Broadband Commission's advocacy Target 2 of 2% affordability threshold – as covered in the previous section.

The challenge I observe in many developing economies is the lack of the national [typically fibre] middle mile backbone, inter-city fibre networks, metro intra-city networks, etc. Take the West Africa Cable System (WACS) cable, which is a submarine communications cable that connects South Africa to the United Kingdom, running along the west coast of Africa. It is owned by a host of African and international fixed and mobile telecoms companies including MTN Group, Angola Cables, Broadband Infraco (South Africa), Cable & Wireless, Congo Telecom, Office Congolais des Postes et Télécommunications (OCPT), PT Comunicações, Togo Telecom, Tata Communications, Telecom Namibia, Telkom SA, and Vodacom [392]. Yet, these same companies as MTN or Vodacom are restricted in many African countries by policy and regulation from building the extensive middle mile national backbones necessary to monetise their first mile investments in cables like WACS. It is also in the commercial interests of big national fixed & mobile telecoms companies to collaborate – Public

---

[389] https://btw.media/company-stories/who-owns-undersea-cables/
[390]Digital Opportunities in African Businesses -
https://documents1.worldbank.org/curated/en/099747205152435810/pdf/IDU1bb3afe0b
1d7f21413b19be21f92001a3b56e.pdf
[391] https://www.itu.int/dms_pub/itu-d/opb/ind/d-ind-ict_mdd-2024-4-pdf-e.pdf
[392] https://businesstech.co.za/news/telecommunications/12582/wacs-ownership-breakdown/

Private Partnership (PPP) style - with Governments, USFs and other private investors on building the middle-mile backbones.

| | Order of Costs | Payback period | Examples |
|---|---|---|---|
| Passive layer | 70–80% of network costs | 15 years | Trenches, ducts, and dark fibre |
| Active infrastructure layer | 20–30% of network costs | 7-year rate of return | Electronic 3G/4G equipment |
| Service layer | N/A | A few months to 3 years | Content, services, and applications |

Table 9.4—Investing In Different Network Layers for Broadband Infrastructure (Toure, 2013)

A PPP for such middle-mile backbones is recommended because of the order of costs of the passive layer, as shown in Table 9.4. The passive layer can account for up to 80 percent of the costs, but has an exceptionally long payback period of approximately fifteen years. In contrast, the active layer, where the intelligence resides, has a five- to seven-year payback period, whilst the service layer's payback period is even shorter at a maximum of three years. Governments and policy makers should note the different investment requirements, and hence different types of investors required, for the typical three layers of broadband infrastructure as depicted in Table 9.4. Governments or states co-financing the passive layer with fixed and mobile operators, along with other long-term private investors on a PPP basis, would make much sense, market conditions permitting.

iii.  *Satellite telecom players should continue developing and providing alternative first-mile Internet connectivity that would reach the 2.6 billion unconnected*: companies like Amazon (through Project Kuiper) and SpaceX (with Starlink) should continue deploying satellite constellations to provide affordable broadband to remote and underserved areas.

## 9.5 *Multilateral / Intergovernmental Leadership*

The role of Multilateral / Intergovernmental leadership with respect to connecting the next 2.6 billion unconnected is just invaluable. I note earlier in this chapter that such leadership includes those from the G7[393], the G20[394], European Commission (EC), the United Nations (UN), the United Nations Development Programme (UNDP)[395], the Commonwealth Telecommunications Organisation (CTO), and leadership from other regional intergovernmental organisations such as the APT[396], the ATU[397], CARICOM[398] and more. I have obviously excluded other intergovernmental organisations like the ITU which I have already covered earlier.

Drawing from this and earlier chapters in this book, here are my top recommendations as regards multilateral and/or intergovernmental leadership to connecting the unconnected 2.6 billion to the Internet:

i.   *Neighbouring and even non-neighbouring developing economies who are part of formal or informal multilateral and/or intergovernmental organisations (e.g. BRICS[399], APT or CARICOM) can provide significant leadership and collaborations in bridging their digital divides and connecting more of their populations to the Internet:* earlier in Chapter 7 (Section 7.3.4), I note how leadership and collaborations amongst the 27 European Union (EU) nations in the domains of telecoms policy, regulation and investments over the past 30 years have undoubtedly benefitted all of

---

[393] The G7 is an intergovernmental forum made up of Canada, France, Germany, Italy, Japan, the UK, and the US. The European Union also participates but is not an official member.

[394] "The G20 comprises 19 countries (Argentina, Australia, Brazil, Canada, China, France, Germany, India, Indonesia, Italy, Japan, Republic of Korea, Mexico, Russia, Saudi Arabia, South Africa, Türkiye, the United Kingdom, and the United States), the European Union, and since 2023, the African Union." – Source: https://g20.org/about-g20/g20-members/.

[395] https://www.undp.org/digital/standards/2-bridge-digital-divide – UNDP was the first in the UN System to adopt a "Digital by Default" approach.

[396] Asia-Pacific Telecommunity – https://www.apt.int/

[397] African Telecommunications Union – https://atuuat.africa/about/

[398] Caribbean Community – https://caricom.org/

[399] https://infobrics.org/ – BRICS is an informal intergovernmental organisation consisting today in 2025 of ten countries – Brazil, Russia, India, China, South Africa, Egypt, Ethiopia, Indonesia, Iran and the United Arab Emirates. The BRIC acronym originated from its first four founding countries: Brazil, Russia, India and China (BRIC).

the countries with the EU. I note how the EU's leadership and collaborative approach driven by the EC has definitely played a significant role in yielding major benefits such as harmonised regulations, and joint investments in digital infrastructure projects, such as the Connecting Europe Facility[400]. The collaborations have led to improved connectivity and digital services across the region, as well as much joint research and innovation across collaborative research, and innovation initiatives such as Horizon Europe[401]. Other multilateral and intergovernmental organisations could and should learn from the EU's experience and adapt similar strategies and leadership – as I describe of the EU/EC in Chapter 7.

ii. *Maximise broader digital divide benefits from multilateral leadership/collaboration*: I have already covered much of these earlier in this book, so I provide just a summary of them here. In order to bridge the digital divide and connect their proportions of the 2.6 billion unconnected, developing economies can truly benefit if they use their membership of the most *relevant and competent* multilateral and/or intergovernmental organisations like the APT, the ATU, SADC[402] or BRICS. I note most 'relevant and competent' because a country like South Africa, which is a member of the ATU, SADC and BRICS, may be able to draw on its membership of BRICS to bridge its digital divide more fruitfully than from its membership of the ATU or SADC. The benefits could include some or all of the following: policy advocacy, policy harmonisation, capacity building, technology sharing, innovation to bridge coverage and usage gaps, infrastructure development, funding and investments, and even public private partnerships amongst public and private entities within the multilateral entity. As covered earlier, the EC leadership and collaboration amongst all EU countries have significantly contributed to bridging the digital divide amongst EU states.

---

[400] https://commission.europa.eu/funding-tenders/find-funding/eu-funding-programmes/connecting-europe-facility_en
[401] *Ibid.*
[402] Southern Africa Development Community – https://www.sadc.int/

## 9.6 Big Tech (e.g. FANGAM/GAMMA) Leadership

Big Tech firms such as Facebook (Meta), Amazon, Netflix, Google, Apple and Microsoft are clearly amongst the biggest and most valuable companies in the world today by market valuation and profitability. These firms arguably have the most commercial interest to see the unconnected 2.6 billion connected sooner than later. Expanding Internet connectivity really aligns with their business interests, as it increases their user base and opens new markets[403].

Big Telco, on the other hand, refers to the Big telecom operators that we all know including Vodafone/Vodacom, Telefonica, Verizon, Deutsche Telekom, British Telecom, Orange, MTN, Airtel, etc.

I add that Big Telco vs. Big Tech is today (in 2025) a hot topic of debate in the ICT/TMT sector with the 'Fair Share' debate[404]. This is in large part because of the pains of a radically evolved new Internet value chain (Nwana, 2022), which has generated humongous positive benefits to consumers and citizens with services like TikTok, Instagram, Google and Facebook – whilst, on the other hand, brought with it numerous negative externalities like Internet scams, fake news, misinformation, disinformation, cybercrime and much more (see Nwana, 2022).

Complaints from Big Telcos that services like Skype, Facebook and Google are *not* paying for use of the Big telco infrastructures they "ride on" are rife, though I emphasise that consumers may be (or are) paying for use of infrastructure anyway. Big Telco operators tend to deliberately omit this important detail in their 'Fair Share' debate arguments. They often cite as evidence the rise and rise of OTT Traffic on Telco Networks as depicted in Figure 9.5.

---

[403] https://www.forbes.com/sites/miriamtuerk/2022/10/17/big-techs-push-to-connect-the-unconnected-signals-growth-for-clean-tech/
[404] https://www.cnbc.com/2023/02/28/big-tech-vs-big-telco-top-eu-official-says-theres-no-battle-over-network-funding.html - Big Tech vs. Big Telco: Top EU official says there's no 'battle' over network funding

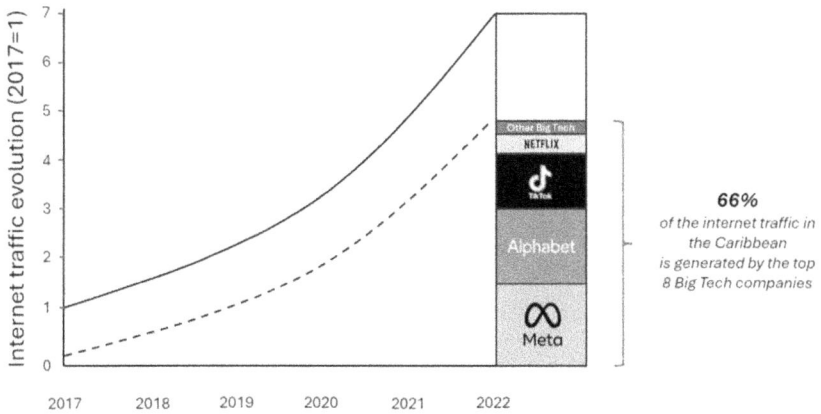

Figure 9.5 – The rise and rise of OTT Traffic on Telco Networks
in the Caribbean between 2017 to 2022
*(Source: Based on and adapted from Digicel Group Presentation [405])*

As shown from this graph (Figure 9.5), 66% of the exponential growth in Internet traffic in the Caribbean is generated by the top 8 Big Tech platforms, with Facebook/Meta, Google/Alphabet and TikTok dominating markedly. The rest of native traditional telco traffic only constitutes 33%. So, Big Telco forcefully asserts that Big Tech should "find a financing model for the huge investments needed" in the development of next-generation mobile/cellular networks, with emerging technologies like the Metaverse[406]. MNOs blame OTTs for declining voice and SMS revenues across the telecoms sector. The reality is that these arguments are more nuanced, and I have covered this in much detail in Nwana (2024), Chapter 6.

---

[405] Source: David Geary of Digicel Group, Presentation at CTO Ministerial Summit, London, Feb 2023, https://www.cto.int/wp-content/uploads/2023/04/CTO-DW23-Event-report.pdf
[406] Source: https://builtin.com/media-gaming/what-is-metaverse – The metaverse is a network of shared, immersive virtual worlds where people can connect with friends, create and play games, work and shop. You can think of the metaverse as a cyberspace, or an evolved, three-dimensional internet where logging in isn't necessary.

To discuss how GAMMA Big Tech could further connect the 2.6 billion unconnected individuals, it is important to first recognise that significant efforts are already being made. However, they can do much more:

a. *Big Tech should continue to build out more first mile submarine cable capacity, particularly into countries with significant unconnected populations:* this is already happening. Earlier, I cite the World Bank study published in July 2024[407] that demonstrates that each doubling of such 'first mile' capacity in Africa generally results in a 7% drop in the price of fixed broadband Internet in Africa, and circa 13% for mobile broadband prices across the African continent. This is more than justification enough.

   For example, Meta's 2Africa project is an ambitious initiative which aims to realise the world's longest submarine cable system, spanning over 50,000 kilometres. This Meta/Facebook cable aims to connect Africa, Europe, and Asia, significantly enhancing Internet connectivity and bringing affordable, high-speed Internet for approximately 3 billion people, or about 36% of the global population [408] [409]. Big Tech content providers such as Google, Meta, Microsoft and Amazon are also major investors in new cables as Telegeography[410] confirms, and they should continue to build, maintain and renew fibre more. Once more I emphasise the fact that there is a dearth of middle-mile fibre backbones to carry data inland, leaving much first-mile fibre capacity significantly under-utilised.

b. *Big Tech should continue being the biggest advocates of affordable Internet connectivity and fund real projects to realise this goal:* companies like Google and Meta/Facebook have launched initiatives like Google Station and Free Basics, respectively, to offer free or low-cost Internet access in developing countries[411]. I have personally witnessed how important some of these basic and free Internet services are in countries like Madagascar (in

---

[407]Digital Opportunities in African Businesses –
https://documents1.worldbank.org/curated/en/099747205152435810/pdf/IDU1bb3afe0b1d7f21413b19be21f92001a3b56e.pdf
[408]https://www.bbc.co.uk/news/articles/ckgrgz8271go#
[409] https://engineering.fb.com/2021/09/28/connectivity/2africa-pearls/
[410] https://submarine-cable-map-2024.telegeography.com/
[411] https://www.forbes.com/sites/miriamtuerk/2022/10/17/big-techs-push-to-connect-the-unconnected-signals-growth-for-clean-tech/

Africa) and in India, particularly when they offer *free* educational, agricultural and health content.

c. *Big Tech has this unmatched capacity to innovate new <u>coverage gap</u> connectivity solutions – and they must continue to do so:* unfortunately, I have observed Big Tech firms like Google, Microsoft and Facebook backing off such connectivity innovations over the last decade to 2025. Big Tech used to invest more in innovative connectivity solutions, such as low-cost antennas and off-grid power systems, innovative low-cost solutions etc., to make connectivity more accessible. Big Tech can productise civilisation-in-a-box (cf. Chapter 4) most effectively. Meta/Facebook used to innovate much a decade ago in lowering telco infrastructure costs as well as lowering the Ownership Business Models for Internet/Voice connectivity (see Tables 3.2 and Figure 3.3, Chapter 2). Sadly, much of such innovations have since stopped. I urge Big Tech to innovate more as it aligns so much more to their commercial interests.

d. *Big Tech also has an unmatched capacity to innovate new <u>usage gap</u> connectivity solutions – and they must innovate these too in a major and collaborative manner:* I have listed before the usage gaps contributors like crushing poverty, poverty (low income), lack of awareness, connectivity being a low-priority concern, illiteracy, lack of digital skills, low population density (rural areas), usage-based pricing models, poverty, miscellaneous issues, etc. – are of these are truly non-trivial to address. So, usage gap digital inclusion programs are direly needed in developing countries.

I have observed in many developing economies several Big Tech companies like Microsoft, Facebook/Meta, Google (and even Big Telcos too) run such programs to equip SMEs, women and communities with devices and digital literacy skills, facilitating that they can fully participate in the digital world. However, much of their efforts are *individually* subscale to the magnitude of the usage gap challenges in these countries. I have always felt if they collaborated and pooled their resources, the impact would be much greater. Perhaps, ICT leadership of these countries should consider leading on this.

e. *Big Tech would have to be more 'sensitive' to Big Telcos' positions on the 'fair share' contentious debate going forward*: Section 9.11 elaborates on this more.

## 9.7  Multilateral Development Bank (MDB) Leadership

The leadership of Multilateral Development Banks (MDBs) like the World Bank Group, the International Finance Corporation (IFC), the Asian Development Bank (ADB), the African Export-Import Bank (AfriExIm Bank), the African Development Bank, etc. is pivotal in bridging the digital divide.

The *World Bank Group*[412] deserves special mention here – which itself consists of five institutions working together to reduce poverty and promote sustainable development globally. The five institutions include the following:

- International Bank for Reconstruction and Development (IBRD): focuses on lending to LMICs and creditworthy LICs for development projects;
- International Development Association (IDA): focuses on providing *concessional loans and grants* to the world's poorest countries to help them address development challenges;
- International Finance Corporation (IFC): focuses on *supporting private sector development* in emerging markets through providing financing and advisory services;
- Multilateral Investment Guarantee Agency (MIGA): focuses on offering *political risk insurance and credit enhancement* to encourage foreign direct investment in developing countries;
- International Centre for Settlement of Investment Disputes (ICSID): focuses on *facilitating arbitration and conciliation* of investment disputes between Governments and foreign investors.

Collectively, these World Bank Group (WBG) institutions aim to reduce poverty, promote sustainable economic growth, and foster shared prosperity globally.

As the reader can infer from the above, the World Bank Group's role in bridging the digital divide more broadly, and connecting the unconnected 2.6 billion in

---

[412] https://www.worldbank.org/ext/en/home

particular, is truly key. So, I use the WBG as a very representative of the MDBs categories. The WBG is very influential across developing countries, including in the following ways:

1. *Digital Policy Advocacy and Reforms*: through advocating for policies that encourage inclusive digital access or promoting public-private partnerships in the ICT sector;
2. *Investments in Digital Infrastructures*, particularly in underserved rural and remote areas;
3. *Promoting Digital Inclusion*, particularly with respect to including women, SMEs and marginalised groups – for them to benefit too from digital transformation;
4. *Digital Trainings and Capacity Building*: this is key to address the usage gaps root causes in particular;
5. *Providing Evidence-Based Insights, Monitoring and Reports,* e.g. the World Bank Market Gap and Access Gap Model (see Figure 8.2), the relatively recent World Bank study published in July 2024[413] that demonstrates that each doubling of such 'first mile' capacity in Africa generally results in a 7% drop in the price of fixed broadband Internet in Africa;
6. *Promoting and Leveraging Innovative Financial Mechanisms*: the WBG can propose the use of blended finance models, using debts instruments like impact bonds and other guarantees to stimulate further private sector capital for digital divide development projects;
7. *Promoting and Fostering Digital Innovation Ecosystems*, including investing in incubation hubs, tech start-ups, digital research and development (R&D) efforts, promoting sandboxes, and more.
8. Etc.

These MDB activities are all invaluable and MDBs should continue such efforts as above.

My key recommendations to MDBs are no different to my five top and 'direct' recommendations on mitigating the digital divide aimed at the leadership of individual LDC-type countries covered in Section 9.1. I believe the WBG should

---

[413]Digital Opportunities in African Businesses -
https://documents1.worldbank.org/curated/en/099747205152435810/pdf/IDU1bb3afe0b1d7f21413b19be21f92001a3b56e.pdf

take these recommendations most seriously and use them to influence and modify their works. In general, I assert that the learnings and insights and recommendations of this entire book would truly inform the WBG's activities in bridging the digital divide.

## 9.8   Charity & NGO Leadership: 'Connect a Billion'

Non-Governmental and Private Charitable Organisations (NGOs), e.g. the Internet Society, A4AI[414], Bill & Melinda Gates Foundation, Tony Blair Institute for Global Change, Vernonburg Group LLC, etc., all contribute to connecting more citizens and bridging the digital divide in developing economies.

I would like to highlight and specifically mention the Vernonburg Group-led, *Connect a Billion* initiative[415]. However, before discussing this further, allow me to provide an overview of several other NGO initiatives focused on connecting the 2.6 billion unconnected.

- *Internet Society*: the Internet Society loudly pronounces that *the Internet is for Everyone*[416]. Since they argue that everyone must have access the "life-changing resource[417]", they advocate and work towards connecting underserved communities using community networks and maintained Internet solutions, which are locally built. The Society promotes and advocates for favourable policies for licensing, spectrum allocation, and public funding to support small-scale connectivity solutions[418]. They also advocate for a "trusted" Internet globally, noting that the world relies on a secure and trustworthy Internet.

- *Alliance for Affordable Internet (A4AI)[419] & the Global Digital Inclusion Partnership[420]*: in Chapter 6, I introduce A4AI's well-researched

---

[414] Alliance for Affordable Internet – www.a4ai.org
[415] https://www.vernonburggroup.com/
[416] https://www.internetsociety.org/action-plan/connecting-the-unconnected/
[417] *Ibid.*
[418] https://www.internetsociety.org/resources/advancing-community-connectivity/
[419] https://a4ai.org/who-we-are/
[420] Global Digital Inclusion Partnership - Meaningful connectivity for the global majority - https://globaldigitalinclusion.org/

*Meaningful Connectivity framework*[421] and how they applied it to nine LMIC economies (Colombia, Ghana, India, Indonesia, Kenya, Mozambique, Nigeria, Rwanda, and South Africa). A4AI has truly been a gem of an organisation that advocates for affordable Internet access by influencing policies and regulations in developing countries. Their recommended policies typically focus on reducing the cost of data and improving infrastructure in developing countries, as well as advocating for improved meaningful connectivity. I have been such a fan of A4AI to date, but I am concerned today in 2025 that USAID – who has been their biggest funder by far – has been dismantled. The dismantling of USAID under the 2nd-term Trump administration has raised significant concerns about the future of organisations like A4AI. This is because, whilst A4AI also receives support from other sponsors like Google and Sweden's SIDA[422], the loss of USAID funding could impact its ability to advocate for affordable Internet access globally. On the positive side, the Global Digital Inclusion Partnership (GDIP) may be emerging to 'buttress' A4AI. The latter's goals are very similar to the A4AI's, not least because the Founder of GDIP is none other than the indefatigable Digital and Gender Inclusion Champion Sonia N Jorge[423] who led the A4AI for close to a decade, up until September 2022.

- *Bill & Melinda Gates Foundation*: they focus on – or have previously focused on – digital connectivity to empower women and girls, particularly in low-income regions[424]. They try to close such digital access gender gaps by supporting action-oriented research, advocacy and innovation.

- *Tony Blair Institute (TBI) for Global Change*: TBI advocates for universal Internet access across specific developing countries and has published reports spelling out practical steps and recommendations to close the digital

---

[421] The framework focuses on four pillars: 4G-like speeds, smartphone ownership, daily use, and unlimited access at a regular location, like home, work, or a place of study.
[422] Swedish International Development Cooperation Agency
[423] https://www.linkedin.com/in/sonia-jorge-38b65711/
[424] https://www.gatesfoundation.org/our-work/programs/gender-equality/digital-connectivity

divide by 2030[425]. TBI emphasises the importance of PPP collaborations amongst the Government, private sector players, and civil society.

- *Vernonburg Group LLC*[426] *and the Connect One Billion Initiative*: the Vernonburg group was founded by a true veteran digital divide champion, Paul Garnett[427], formerly of Microsoft. With support from USAID (principally), Microsoft and Stadia Capital, Vernonburg Group was charged to establish and deploy a Connect One Billion entity for small to mid-sized Internet Service Providers (ISPs) in low and middle-income countries (LMICs), working in consultation with industry and other interagency partners[428]. Before the election of Donald Trump to his second term in November 2024, USAID, Microsoft, Vernonburg Group, and Stadia Capital had spent more than a year co-creating a blended finance entity designed to support and accelerate the growth of ISPs in LMICs.

Connect One Billion's mission is to provide small, high-potential ISPs in LMICs with the capacity building and early financial support they need to extend secure and affordable high-speed Internet connectivity. This way, they aim to connect one billion people over the next decade to the Internet. Connect One Billion would/will be there to support these ISPs at various stages of development and it was envisioned to set up an initial $250M fund initiative with targets of circa $50M in Technical Assistance, $50-100M in Concessional Capital, and $100-150M in Growth Equity.

The idea is/was to have Connect One Billion as the legal entity with its own management, board, and staff; with Vernonburg Group effectively incubating the formation of Connect One Billion, pursuant to an agreement with USAID[429].

I cannot be more effusive about how innovative this Connect One Billion initiative is, and the reader would understand this given how much I laud the fact that Brazil has over 20,000 ISPs but only 3 have country-wide

---

[425] https://institute.global/insights/tech-and-digitalisation/state-of-compute-access-how-to-bridge-the-new-digital-divide
[426] https://www.vernonburggroup.com/
[427] https://www.linkedin.com/in/paul-garnett-32403ba
[428] https://www.highergov.com/grant/7200AA24CA00012/
[429] https://www.highergov.com/grant/7200AA24CA00012/

coverage – with 40% of the companies possessing up to 5 thousand customers (cf. Section 5.3, Chapter 5). I strongly argue in that section that democratising and localising connecting the next 2.6 billion through high-potential ISPs is exactly what is needed to connect more of the unconnected 2.6 billion unconnected to the Internet today.

I desperately hope that the Trump 2.0 Administration does not kill off this brilliant initiative through the dismantling of USAID. I declare an interest that Paul Garnett is a friend who I have worked closely with before at the Dynamic Spectrum Alliance[430] which he largely founded – but this Connect a Billion initiative is exactly the sort of innovation needed to make a significant dent to the 2.6 billion unconnected.

This aforementioned non-exhaustive list of NGOs all contribute to addressing digital inequality globally. However, the most impressive NGO effort I have observed in our industry is undoubtedly the Connect a Billion initiative.

## 9.9   *Government Aid & Development Leadership*

I have personally observed the difference that some individual Governmental Aid programmes can make in some developing countries. For example, the Australian Government DFAT's[431] Economic and Social Infrastructure Program (ESIP) is indeed making a significant impact in LMIC Papua New Guinea (PNG) including addressing digital divide challenges. This ESIP program focuses on improving infrastructure investment and planning, procurement, financing, and delivery in key sectors like telecommunications, energy, health, water and education. Recall from Section 2.2 that PNG – which has a circa 85% rural population – has a "two-tiered economy": one tier is occupied by about 70 per cent of the population who live at a subsistence level; and the other tier has fewer than 250,000 people who are formally employed, i.e. circa one in 25. The Australian ESIP programme prioritises high-quality and sustainable infrastructure investments in PNG in order to enhance the quality of life for this 70% poor citizenry across PNG. The program also supports initiatives that

---

[430] https://www.dynamicspectrumalliance.org/
[431] https://www.dfat.gov.au/publications/development/pngs-economic-and-social-infrastructure-program-esip

strengthen PNG's capacity to manage and deliver infrastructure projects effectively. This includes fostering partnerships between the Australian and PNG governments to ensure long-term development goals are met. Australia is evidently one of the largest developmental partners to PNG.

So, individual Governmental Aid programmes like the Australian Government's [DFAT] ESIP programme or the UK's Foreign, Commonwealth & Development Office (FCDO) play a pivotal role in addressing the digital divide and connecting the 2.6 billion unconnected. The leadership of Governmental Aid programmes typically focus on the following similar areas (as with the leadership of MDBs of the previous section), namely: Digital Policy Advocacy and Reforms, Targeted Funding for Digital Inclusion, Investments in Digital Infrastructures, Promoting Digital Inclusion, Digital Trainings and Capacity Building, Providing Evidence-Based Insights, Monitoring and Reports, Promoting and Fostering Digital Innovation Ecosystems, etc. These Governmental Aid-led activities are all invaluable and should continue such efforts as above.

My key recommendations to Governmental Aid leadership to help address the 2.6 billion not online are also no different to my five top 'direct' recommendations aimed at the leadership of individual LDC-type countries covered in Section 9.1. I believe Governmental Aid programmes should take these recommendations most seriously and use them to influence and modify their activities as regards addressing the digital divide.

I would just add one more specific recommendation here with respect to Governmental Aid leadership.

*Collaborations amongst the Governmental Aid agencies and with specific MDBs are strongly recommended:* Governmental Aid bodies like USAID (being dismantled by the Trump 2.0 Administration), UK Aid, Australia Aid/DFAT and so many more[432] can and should collaborate amongst themselves

---

[432] A longer more comprehensive (non-exhaustive) list includes (in no particular order):
1. African Development Bank (AfDB)
2. World Bank – International Development Agency (IDA)
3. Inter-American Development Bank (IaDB)
4. Asian Development Bank (ADB)
5. US Millennium Challenge Corporation (MCC)
6. U.S. Agency for International Development (USAID)

and work alongside Multilateral Development Banks (MDBs) and other international organisations to *amplify* their developmental impacts.

I am quite aware that each sovereign Government has their individual and *soft power* reasons to provide much-needed developmental aid to specific countries, e.g. Australian Aid/DFAT is a major developmental partner to neighbouring

---

7. United Nations Children's Fund (UNICEF)
8. United Nations Development Programme (UNDP)
9. Bundesministerium für wirtschaftliche Zusammenarbeit (BMZ-GIZ). Deutsche Gesellschaft für Internationale Zusammenarbeit (GIZ) is the German development agency that implements projects on behalf of the German government, including BMZ.
10. China - Ministry of Commerce of the People's Republic of China (MOFCOM)
11. Korea International Cooperation Agency (KOICA)
12. European Commission (EC) Directorate-General for International Partnerships (INTPA) - a department within the EC responsible for formulating the EU's international partnership and development policy. Its mission is to reduce poverty, ensure sustainable development.
13. United Kingdom - Foreign, Commonwealth & Development Office (FCDO)
14. France - Agence Française de Développement (AFD)
15. Canada - Global Affairs Canada (GAC)
16. Sweden - Swedish International Development Cooperation Agency
17. Japan - Japan International Cooperation Agency (JICA)
18. New Zealand - Ministry of Foreign Affairs and Trade (MFAT)
19. Gates Foundation
20. European Bank for Reconstruction and Development (EBRD)
21. European Investment Bank (EIB)
22. Italy - Italian Agency for Development Cooperation (AICS)
23. World Bank/International Finance Corporation (IFC)
24. Ireland – Irish Aid
25. Switzerland - Swiss Agency for Development and Cooperation (SDC)
26. Australia - Department of Foreign Affairs and Trade (DFAT)
27. Spain - Spanish Agency for International Development Cooperation (AECID), or Agencia Española de Cooperación Internacional para el Desarrollo
28. Saudi Arabia – Kingdom of Saudi Arabia Relief (KSRelief)
29. IDB Invest - IDB Invest is the private sector arm of the Inter-American Development Bank (IDB)
30. Turkey - Turkish Cooperation and Coordination Agency (TIKA)
31. UAE - Ministry of Foreign Affairs and International Cooperation (MOFAIC)

PNG and Pacific SIDS. Similarly, the UK's Foreign, Commonwealth & Development Office (FCDO) has been focusing on countries like Nigeria, South Africa, Indonesia, Brazil, and Kenya through initiatives such as its Digital Access Programme[433]. This UK programme aims to promote inclusive, affordable, and secure digital access for underserved communities in these nations. My point here is there are likely overlaps in developed countries supporting digital developmental programmes in developing countries. For example, Australia is also a significant developmental partner to Indonesia[434], as is the UK. So, joint initiatives between Australia, the UK and other MDBs can ensure a more coordinated approach to bridging the digital divide and connecting more Indonesians to the Internet in Indonesia.

## 9.10 Can Connectivity Leadership Groups Work Together More Effectively?

I genuinely do *not* ask this question rhetorically. I hope it is clear from my overview of the non-exhaustive list of categories of digital connectivity leadership across the globe that it can all be very confusing. I also know this because some Ministers, senior regulators and even CxOs of telecoms businesses from developing countries have asked us/me how to navigate this connecting-the-unconnected ecosystem during consultancy engagements or trainings with our consultancy, Cenerva[435].

I do *not* have the answer, but it appears to me that some coordination amongst all these leadership categories would help developing economies receive more developmental impacts. As I explain at the end of the previous section, a joint coordinated approach and initiative between Australia, the UK and other MDBs in Indonesia, say, would surely deliver greater connectivity impacts in Indonesia. I have personally worked in developing countries like Togo (in West Africa) and have observed the clear need for better coordination of efforts from UN agencies, the World Bank Group entities, the African Development Bank, Big Tech, Big Telco, etc. on digital connectivity matters. How this is realised is something that is beyond the scope of this chapter.

---

[433] https://devtracker.fcdo.gov.uk/programme/GB-1-204963/summary
[434] https://www.dfat.gov.au/geo/indonesia/development-assistance/australia-partnership-indonesia
[435] www.cenerva.com

## 9.11 Financing Broadband for the Unconnected – 'Big Tech's Digital Colonialism'

I decided to add the last core section of this chapter both in the interest of good scholarship and also because it squarely concerns the 'Fair Share'[436] or "fair contribution"[437] debates that I briefly introduce earlier in this chapter (see Section 9.6).

During 2024, the ITU estimated that USD 1.6 trillion is required in *hardware costs alone* to complete the task of connecting the world, with most of the resources needed in developing countries[438]. This is the estimate required to cover the costs of building networks, installing necessary equipment, expanding access to underserved regions, and to bridge the divide in a sustainable and efficient manner. The ITU also emphasises the importance of partnerships and innovative solutions, like satellite technology and public-private collaborations.

However, according to the well-known Danish Consultancy *Strand Consult*[439], the world lacks some USD $2 trillion in network investments to cover the 2.6 billion third of the world's population that remains offline. Strand Consult rightly observes that the business models used to recover network investment costs are incredibly challenging for this investment, particularly in emerging countries with high costs for capital and energy. This much is *not* controversial at all.

Strand Consult proceeds to make the case the mobile operators business models could recover their network investment costs in the days before video traffic began to predominate. Strand Consult asserts that Big Telco network providers experience that the cost to manage video traffic in their networks grows faster than revenue, whilst they cannot raise prices to manage cost (see Figure 9.5

---

[436] https://www.cnbc.com/2023/02/28/big-tech-vs-big-telco-top-eu-official-says-theres-no-battle-over-network-funding.html

[437] https://connecteurope.org/insights/blog/9-questions-and-answers-fair-contribution-debate

[438] https://www.itu.int/en/mediacentre/Pages/PR-2025-03-04-Mobile-World-Congress.aspx#

[439] https://strandconsult.dk/gigabit-caribbean-closing-the-investment-gap-in-fixed-and-mobile-networks/

about the Caribbean OTT traffic on Big Telco networks). Strand proceeds to call out the Big Tech 'abuse' of "nations and telecoms operators around the world" [440].

These challenges are described in part in Strand Consult's research published in a September 2024 research note <u>provocatively</u> entitled *The Caribbean is a microcosm of **Big Tech's digital colonialism**. Small and medium-sized emerging countries are profitable to exploit*[441]. The UN classifies and categorises the Caribbean as Small Island Developing States (SIDS). Here are some of the conclusions in the Strand Consult Caribbean research note:

- 15M people of the Caribbean states are offline and not connected. The Caribbean consists of some two dozen SIDS nations and 45 million people. So, one-third of the Caribbean are unconnected to the Internet as of early 2025. [Interestingly, the 2025 GSMA Mobile Economy Latin America report[442] found that 22.8 million Caribbean residents – not 15 million – are still offline, 2 million due to coverage gap and 20.8 million due to usage gap challenges].
- Strand Consult calculates the investment gap to realise the Gigabit Caribbean Society at circa USD $8 billion – 13 billion. The concept of the Caribbean Gigabit Society aims at the [potential] transformative role of universal connectivity in shaping the region's future. By setting a benchmark of 100 Mbps for all, it lays the foundation for greater access to digital tools and services, which can drive employment opportunities, innovation, and economic growth.
- Strand Consult calculates that the Caribbean mobile telecom and broadband industry earn annual revenue of more than USD $2.6 billion. More importantly, it notes that they are registered in the region, "pay the local tax,

---

[440] https://strandconsult.dk/the-caribbean-demonstrates-how-big-tech-abuses-nations-and-telecom-operators-around-the-world/

[441] https://strandconsult.dk/the-caribbean-is-microcosm-of-big-techs-digital-colonialism-small-and-medium-sized-emerging-countries-are-profitable-to-exploit/

[442] https://www.gsma.com/about-us/regions/latin-america/a-call-to-action-to-ensure-the-development-of-the-internet-and-the-digital-future-of-latin-america-and-the-caribbean/

apply for regulatory and licensing fees, employ local people, and invest 18% of revenue in infrastructure in the Caribbean"[443].

- However, "Big Tech firms Alphabet, Meta, Apple, Amazon, Microsoft, and Netflix earn in excess of $11.5 billion in the Caribbean yet pay no tax, no registration, no regulatory or license fees, nor local salaries, or local investment"[444].

- Strand Consult observes that Broadband Telco providers make significant investments every year in the Caribbean whilst "Big Tech uses Caribbean networks while paying little to nothing, the textbook example of a free ride"[445].

Strand Consult concludes that *"Big Tech's **digital colonialism** undermines the economics for the Caribbean broadband telecom and broadcast/media industries. Local players support the local economy while Big Tech free rides"*[446] (*the author's emphases*). The 'abuse of nations and telecoms operators' and 'Big Tech's digital colonialism' say it all in summary – and the reader can glean from this section the fervour of the 'fair share' or 'fair contribution' debates. The term 'fair contribution[447]' refers to an EU policy discussion regarding whether technology companies should contribute to the funding of gigabit networks. This debate commenced in 2022 with the European Declaration on Digital Rights and Principles, which mandated that "all market actors" should equitably contribute to infrastructure costs.

My personal take on this debate is both different and more nuanced than Strand Consult's, and I cover this in Nwana (2022, Chapter 6), in a section entitled *Briefly Busting Some OTT Myths*. Nevertheless, I believe it is important for the audiences of this book to be aware of this contentious debate between Big Tech and Big Telco – and indeed LIC nations – as regards financing broadband Internet networks for the unconnected 2.6 billion. I am sympathetic to the position that if it would indeed cost USD $2 trillion in network investments to

---

[443] https://strandconsult.dk/the-caribbean-is-microcosm-of-big-techs-digital-colonialism-small-and-medium-sized-emerging-countries-are-profitable-to-exploit/
[444] *Ibid.*
[445] *Ibid.*
[446] *Ibid.*
[447] https://connecteurope.org/insights/blog/9-questions-and-answers-fair-contribution-debate

cover the 2.6 billion unconnected, and I do *not* believe the current Big Telco industry would foot all these monies out of their current business models.

In this vein of raising the necessary funds to address global connectivity, I note earlier in this chapter (Section 9.2) that the ITU has called for USD 100 billion in commitments to its Partner2Connect (P2C) Digital Coalition in order to drive digital transformation across all countries, with special attention to the hardest-to-connect communities, particularly in Least LDCs, LLDCs, and SIDS[448]. The ITU announced at Mobile World Congress in March 2025 that its P2C Digital Coalition has reached an "impressive" $73 billion in pledges[449] towards 'universal, meaningful connectivity'. The ITU hailed the growing momentum from leading mobile and satellite industry players rallying to bridge the digital divide, citing new commitments from MTN Group and GlobalStar[450].
ITU SG Doreen Bogdan-Martin noted that:

> "these new pledges bring us even closer to our goal of delivering sustainable digital transformation to everyone, everywhere"[451].

Really? As I note earlier in this section, the ITU itself estimates USD 1.6 Trillion in hardware costs alone to cover the costs of building networks, installing necessary equipment, and expanding access to underserved regions. But the ITU's P2C Digital Coalition fund aims to raise USD 100 billion to advance universal and meaningful connectivity worldwide. I applaud the ITU's P2C efforts, but the 2.6 billion unconnected will *not* be meaningfully connected through ITU's P2C initiative. So, the 'fair share' debate lives on how a minimum of USD 1.6 Trillion would be raised.

## 9.12 Summary

This chapter overviews the many categories of leadership that are all involved in some way in bridging the connectivity digital divide across the globe, not least because it gets quite confusing – and I have observed some countries 'playing'

---

[448] https://www.itu.int/partner2connect/
[449] https://broadcastmediaafrica.com/2025/03/05/itus-digital-coalition-secures-us73-billion-in-connectivity-commitments/
[450] https://www.itu.int/partner2connect/p2c-flash-edition-02/
[451] https://singaporeupdate.com/partner2connect-digital-coalition-reaches-73-billion-in-pledges/

off one or more category/categories of 'leadership' players against others. This chapter makes recommendations across all these categories drawing from the insights and evidence in this book. I end this chapter with some sense of the scale of the financial investments required to connect the unconnected 2.6 billion, and the related 'fair share' debate on who should fund these much-need monies.

# 10  The Book Summarised

"A problem well stated is half solved."
— Charles Kettering, Head of Research at
General Motors from 1920 to 1947

One key piece of valuable advice from a peer reviewer[452] of this book was to create a completely standalone summary chapter, "largely independent" of the main book. He advised treating the reader as unfamiliar with the previous nine chapters when writing this summary. So, I have followed this reviewer's suggestion and thoroughly rewritten this final chapter from what I had before, and I have followed his proposed structure:

1.  *What* are the main digital connectivity challenges and why they have not been successfully addressed to date?
2.  *What* are the prospects of the connectivity challenges being addressed in short order, e.g. by 2030? They will not.
3.  *What* kinds of technical solutions are needed?
4.  *What* key warnings should connectivity leaders of LDCs heed?
5.  *Why* the ITU must take meaningful connectivity seriously?
6.  *What* happens when connectivity markets fall short? Learnings from India.
7.  *What* are some recommended <u>initial</u> and <u>minimal</u> clear action points and timescales, preferably in a table showing what's needed and who is responsible for doing it (and ideally why it will work this time round)?

## 10.1  The Connectivity Crisis: Half the World Left Behind

I profess in this book that we have a true *connectivity crisis* on our hands. I am a big fan of the Charles Kettering quote that I cite right at the beginning of this chapter: "a problem well stated is half solved'. So, I think it is imperative that I state and define the connectivity crisis most clearly.

My bold assertion of the crisis stems from some key stubborn facts and evidence.

---

[452] I thank Prof William Webb for this sage advice.

a. Firstly, that the universal digital inclusion challenge is nowhere close to being realised for 2.6 billion people who remain *completely* unconnected as of 2025. This is according to the International Telecommunications Union's (ITU's) Facts and Figures 2024 report[453]. The distribution across regions of the 2.6 billion unconnected to the Internet would not be that different to what it was in the ITU's 2022 Facts & Figures[454]: the majority of the unconnected reside in Africa (60%), Asia-Pacific (36%), the Arab States (39%), the Americas (17%), CIS (16%), and Europe (11%).

b. Further, I estimate in this book using ITU, A4AI and GSMA data that there are more than 2 billion more globally who are connected to the Internet, but who hardly experience any 'meaningful connectivity' at all. I assert this firmly because the ITU's approach and manual for measuring ICT and Internet access use[455] [by households and individuals] considers accessing the Internet *once* "in the last 3 months" to be *meaningful connectivity*. I consider this is an incredibly minimal measure that does not necessarily, in my opinion, reflect meaningful connectivity. I strongly argue in this book that meaningful connectivity – by my definition – implies regular, dependable, and effective access to the Internet – one that enables the users to benefit from online resources and services anytime, anyplace and anywhere. This hardly translates to once in three months.

2.6 billion *unconnected* and a further 2 billion *under connected*. This makes more than half of humanity unconnected or under-connected, hence the title of this book *The Connectivity Crisis - Half the World Left Behind*. Figure 10.1 broadly depicts these challenges.

c. Furthermore, in this book, I conservatively estimate that between 1,294 million and 2,000 million people aged over ten years *may* be unable to *regularly* make a basic phone call by 2025. However, I acknowledge that a

---

[453] *Ibid.*

[454] https://www.itu.int/itu-d/reports/statistics/facts-figures-2022/index/

[455] *Manual for measuring ICT access and use by households and individuals – 2020 Edition* - https://www.itu.int/dms_pub/itu-d/opb/ind/D-IND-ITCMEAS-2020-PDF-E.pdf

significant portion of these individuals may have access to a basic phone device through other members of their households.

In clear terms, the telecommunications market has been unsuccessful in providing *basic voice services* to over one billion (potentially up to two billion) people worldwide. This is the grim reality in 2024/25, thirty-three (33) years after the first 2G network and GSM call was made in Finland[456].

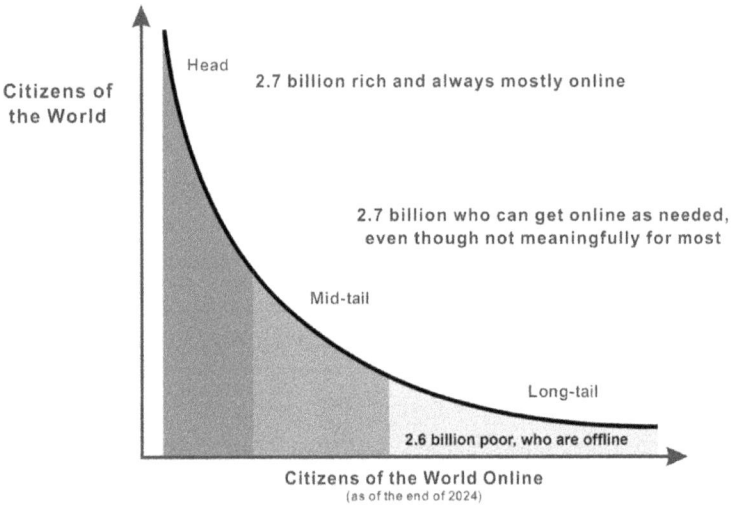

Figure 10.1[457] – The Long Tail Model[458] - the *head* "2.7 billion rich and always mostly online," the *mid-tail* " 2.7 billion who can get online as needed, even though not meaningfully for most" and *the tail* "2.6 billion poor[459] who are *offline* in 2025."

---

[456] https://www.gsma.com/about-us/who-we-are/our-history/
[457] Same as Figure 1.1 in Chapter 1, with no apologies for repeating in this final chapter. This is largely because I assume the reader has not engaged the rest of the preceding main chapters of this book at all.
[458] Source: https://medium.com/@jasper_39476/how-will-long-tail-affect-the-leisure-branch-dec45338a5a - How will long tail affect the leisure branch? | by Jasper Dik | Medium
[459] 'Poor' as defined by World Bank and UN developmental metrics and nomenclatures, e.g. LDCs, LICs, LMICs, etc.

## 10.2 Why the Connectivity Crisis Won't Be Fixed Fast – Insights from Pareto

This book outlines the main reasons why the 'Tail 2.6 billion' are not connected online in 2025 *at all* (see Figure 10.1). It also discusses why a significant portion of the 'Mid-Tail 2.7 billion' are not *meaningfully connected* – I assert up to 2 billion. What are the prospects of this connectivity crisis being resolved soon, such as by 2030 – as per the United Nations (UN) 2030 Agenda for sustainable development[460]? Sadly, the likelihood appears to be very low indeed.

The primary reasons I find in this book fall into three broad categories: (i) demand-side market failures due to *usage gaps*, (ii) supply-side *coverage gap* market failures, and (iii) policy, regulatory, and leadership shortcomings. Indeed, in many countries, there is a combination of all three categories.

Usage gaps refer to populations that live within the footprint of a voice and/or broadband/Internet network but do *not* use voice and/or the Internet. Coverage gaps refer to populations that do *not* live within the footprint of a voice or broadband network at all. Of the two gaps, the more pernicious – by far – is the usage gap challenge. Authoritative research and surveys such as the University of Cape Town's well-thought out *After Access Surveys[461]* demonstrate unequivocally that the main barriers to digital access and use are *demand-side usage gap* challenges, not *supply-side coverage* issues. Their surveys show and confirm what I profess in the book – that digital inequality is not just about infrastructure but *even more* about affordability, digital literacy, and socio-economic factors[462]. Yet the ITU and everyone else follows the ITU lead and continue to spend millions of dollars on supply-side digital divide solutions.

To illustrate the three categories of reasons behind the connectivity crisis, I discuss two case studies from Papua New Guinea (Southeast Asia) and Malawi (Southern Africa). These studies highlight the real-world challenges of achieving universal digital inclusion and meaningful connectivity in these countries. It is highly unlikely that all citizens of PNG or Malawi will be digitally

---

[460] https://sdgs.un.org/2030agenda
[461] https://researchictafrica.net/project/after-access-2022-survey/
[462] https://afrisig.org/sites/default/files/pdf/Acess-After-Access-AfriSIG-2024.pdf

connected, let alone meaningfully connected, by 2030. So, I assert that achieving the United Nations (UN) 2030 Agenda and the UN's own 17 Advocacy Targets[463] in most LDCs are patently unrealistic.

In this book, I go further by drawing valuable insights from the Pareto principle. Pareto's *law of imbalance* was first realised and made famous in 1896 by Italian economist Vilfredo Pareto[464] when he realised the 'imbalance' that approximately "80% of the land in Italy was owned by only 20% of the population". He quickly realised that this 80/20 rule seemed to be a *universal heuristic* or rule that could be applied to practically all aspects of life. Figure 9.3 shows several Pareto curves: 80:20, 90:10, 70:30, 60:50, etc.

I acknowledge in this book that the Pareto principle is not scientific QED[465], but a rule of thumb that seems to always hold. I write this to address the rightfully sceptical reader at this juncture. I assert that Pareto provides useful and realistic predictions of the timescales and effort levels to realise the solutions to the connectivity crisis. I have drawn from the Pareto principle (80/20 curve) to bluntly assert that 86% of the effort to lift the 2.6 billion into online connectivity *is still ahead of us* – see Section 1.3.1 on how this figure is derived. This prediction clearly accords with the much harder nature of the demand-side, usage gap challenges that afflict developing countries. So, drawing from insights like the Pareto principle and the realities on the ground in countries like PNG and Malawi, I state clearly in this book the long-haul nature of the connectivity challenges to developing economies because of this 86% of load effort still being ahead of us.

I also warn in this book that the telecoms market and industry – as we know them today – would *not* address these connectivity crisis challenges because there are clear Pareto 'limits' to the commercially minded and rational [mobile/fixed] telecoms industry. The reader may wonder why I use this phraseology of Pareto limits in this context. Consider that I read in March 2025

---

[463] https://www.sightsavers.org/policy-and-advocacy/global-goals/ - What are the Sustainable Development Goals? | Policy and advocacy | Sightsavers

[464] https://www.britannica.com/money/Vilfredo-Pareto

[465] Q.E.D. stands for the Latin phrase *quod erat demonstrandum,* literally translated as "that which was to be demonstrated", or - more clearly as - "Just what we set out to prove".

an interview that Bharti Airtel CEO Gopal Vittal gave in which he insisted that the mobile/cellular operator *had made a "very brave call" in deciding to go after 40 per cent of the customers which account for almost 80 per cent of revenue*[466]. This honest CEO said the quiet part out loud. So, I draw the attention of you the reader back to Figure 1.3 (Chapter 1) – the Pareto Distribution Curves. I observe that if the reader studies the curves carefully, he/she will note that the '40% of subscribers generating 80%' of revenues' maps or falls closely to the 70/30 Pareto curve, i.e. the curve also predicts that 30% of the subscribers generate 70% of the revenues.

Look again and the 70/30 curve also predicts that 60% of the Vittal's Airtel subscribers would broadly generate 90% of their revenues. So, bluntly, the last 40% of the subscribers (from 60%-100%) only generate 10% of the revenues. I posit that Bharti Airtel cares 'less' about these latter 40% of their subscribers, and this is me saying this diplomatically. Mobile/cellular companies know these Pareto numbers, and they dictate what rational profit-maximising mobile/cellular operators would do. This is what I refer to as the 'Pareto limits' for mobile/cellular operators (see also Figure 7.2).

Furthermore, the ITU estimates that USD 1.6 trillion is required in hardware costs *alone* to cover the costs of building networks, installing necessary equipment, and expanding access to underserved regions in developing countries, to bridge the divide in a sustainable and efficient manner[467]. Just as this book was going into final publication, the ITU published their September 2025 'Connecting humanity action blueprint: advancing sustainable, affordable and innovative solutions' report[468]. The report estimates that the world needs between USD 2.6 trillion and USD 2.8 trillion to connect humanity by 2030: USD 1.5-1.7 trillion for digital infrastructure, USD 983 billion for affordability, USD 152 billion for digital skills and USD 600 million for policy and regulation. I warn that these monies, particularly the digital infrastructure USD trillions, will

---

[466] https://www.mobileworldlive.com/airtel/airtel-ceo-underscores-strategy-shift-for-turnaround/ OR https://www.mobileworldlive.com/airtel/airtel-ceo-underscores-strategy-shift-for-turnaround/ AND https://www.mobileworldlive.com/old_latest-stories/interview-bharti-airtel-group/
[467] https://www.itu.int/en/mediacentre/Pages/PR-2025-03-04-Mobile-World-Congress.aspx#
[468] https://www.itu.int/dms_pub/itu-s/opb/gen/s-gen-invest.con-2025-pdf-e.pdf

*not* be very forthcoming from an industry already embroiled in a profound 'fair share' or 'fair contribution' debates that I describe in Sections 9.6 and 9.11. "Fair contribution[469]" is an EU policy debate on whether tech companies should help fund gigabit networks. It began in 2022 with the European Declaration on Digital Rights and Principles, which stated that "all market actors" should contribute fairly to infrastructure costs.

Whatever you may think about these debates, the positions of Big Telco and Big Tech are well "dug in" – and this helps no-one in the unconnected 2.6 billion Tail or meaningfully connecting another 2 billion more.

## 10.3 What Kinds of Technical and Regulatory Solutions?

The last section showed how insights from Pareto and the realities on the ground in countries like PNG and Malawi prove the long-haul nature of the connectivity crisis in developing economies, not least because Pareto predicts that 86% of load effort still being ahead of us. Such a load 'lift' still being ahead of us (see Figure 10.2) clearly needs 'out-of-the-box' thinking, insights and other innovative solutions.

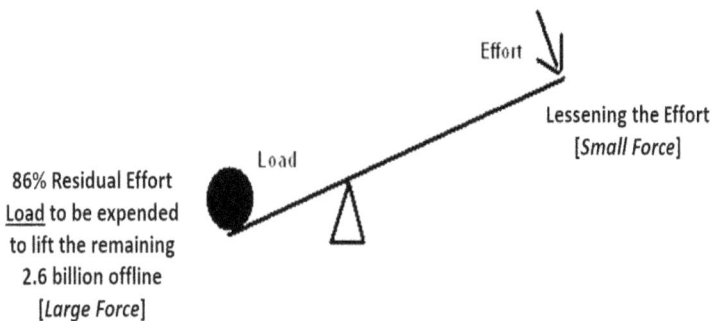

Effort

Lessening the Effort
[Small Force]

Load

86% Residual Effort
Load to be expended
to lift the remaining
2.6 billion offline
[Large Force]

Figure 10.2 – How to Lessen the Large 2.6 billion (86%) Load of the Long Tail Unconnected[470] – as predicted the Pareto 80/20 rule

---

[469] https://connecteurope.org/insights/blog/9-questions-and-answers-fair-contribution-debate

[470] Same as Figure 1.4 in Chapter 1, with no apologies for repeating in this final chapter.

In this book, I derive insights from fundamental concepts such as lever and pulley systems, as illustrated in Figure 10.3. It is recommended that various solutions with 'lever' and 'pulley' characteristics are employed to alleviate the substantial effort required to address the 86%, 2.6 billion load lift challenge, thereby reducing the time needed to achieve successful outcomes (refer to Figures 10.2 and 10.3). The use of additional 'pulleys' decreases the necessary effort (Figure 10.3). The significant challenge posed by the 2.6 billion individuals lacking connectivity requires solutions and approaches embodying such lever and pulley principles.

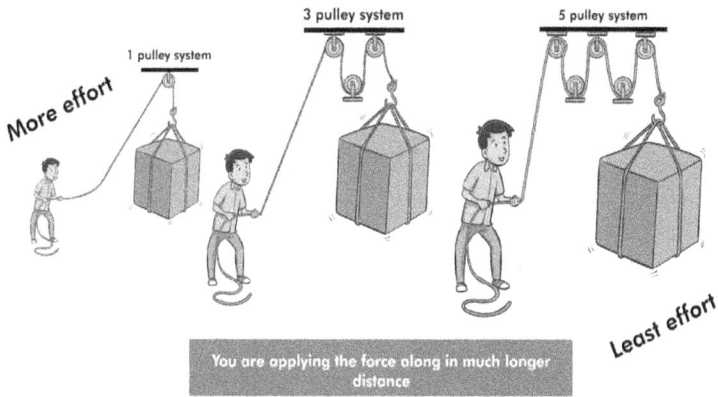

Figure 10.3 – Pulley System Characteristics [471] (Adapted from Source: [472])

In this vein, I present a proposal which I call 'civilisation-in-a-box' proposal that I propose and present in Chapter 4. I submit that it exhibits such 'pulley' characteristics. Without such more 'efficient' solutions, it will be exceedingly difficult and time-consuming to connect the remaining 2.6 billion unconnected individuals (see Figure 10.1) in short order. I would love to see the likes of the Broadband Commission – being a true PPP (that includes the involvements of academia, top industry and governments) – drive to facilitate and productise 'innovations' like Civilisation-in-a-box, and other ideas that are even better.

I also argue in this book for more *pro-unconnected* technical and technology solutions and approaches. Such solutions veer away from the inescapable march

---

[471] Same as Figure 3.2 in Chapter 3
[472] https://images.app.goo.gl/6E6v8umvuxnELjwRA

by the telecoms industry to 6G from 5G – a march which is only all about giving more capacity and goodies to those of us in the Head "2.7 billion rich and always mostly online" (see Figure 10.1).

Alternatively, I argue in this book for the mobile telecoms industry to revert to the wildly successful 4G/LTE standard, take out some elements like VoLTE, add seamless Wi-Fi integration, add seamless roaming and include/prioritise sub-1GHz spectrums to realise a 'new', revised *pro-unconnected* mobile standard. This may read as controversial – but really it should not. Indeed, I argue in this book that India has the capability and capacity to make this happen. I assert this boldly because the mobile industry's 5G to 6G trajectory is set and unstoppable – and only a major population country (1.4 billion+) with the requisite technical capabilities can drive such a different *mobile industrial policy*. After all, I point out that on 23rd August 2023, India became only the 4th country to reach the moon by successfully landing its Vikram lander spacecraft near lunar South Pole[473]. At the very least, I want to see debates emerge on new *pro-unconnected* mobile standards.

I argue for the use of widespread and shared technology solutions. Brazilian WISPs have employed *unlicensed* [open] radio spectrum that allows for easier deployment of non-proprietary and unlicensed Wi-Fi (i.e. widespread) networks. 16.37% of Brazilian households in 2023 accessed broadband *by sharing a fixed broadband connection* with a neighbour (see Table 5.1). This is both clever and economically efficient, 'home grown' and democratic too.

So, bluntly, I point out that much touted technologies such as 5G, 6G and LEO satellite constellation solutions are *not* that relevant to the 2.6 billion unconnected (see Section 9.4) – *not least* because of the usage-gaps-dominant nature of the why the 2.6 billion are unconnected. However, to the extent that 5G, LEOs and other innovative technologies are involved, they will remain very niche solutions to the 2.6 billion unconnected.

I also strongly argue in this book for democratising and localising more of the 2.6 billion connectivity challenge. This is because I argue that connecting most of the tail 2.6 billion unconnected will, require from LMICs and LICs, true 'home grown' solutions – e.g., civilisation-in-a-box. So, Governments and

---

[473] https://www.bbc.co.uk/news/world-asia-india-66594520

regulators need to 'democratise' [as much as possible] and localise much policy, regulatory, financing, technical challenges and more. Most importantly, they must also democratise *responsibility* – to *local* and *non-traditional* 'telecoms' stakeholders in society including community NGOs, local schools, local universities, local churches, local supermarkets, local businessmen and businesswomen, etc. Essentially, we would also need to evolve and license a new 'breed' of Internet Service Providers (ISPs) to help connect the poor in these LIC, LMIC and LDC economies. The current traditional 'big boy' MNOs and fixed operators have patently failed to do so and will continue to fail.

In case the reader thinks of the latter paragraph as rather fanciful, Brazil has over 20,000 ISPs today in 2025, but only 3 have country-wide coverage – with 40% of the WISP companies having up to 5 thousand customers (cf. Section 5.3, Chapter 5). Incredibly, their regional ISPs accounted for over 50% market share in Brazil as of December 2022. This is a highly *democratic* [20,000 ISPs with less than 5,000 customers each], competitive and heterogenous marketplace. The 3 'big boy' MNOs have less than 50% of the Internet connectivity market. Enabling this extremely high number of ISPs has been a real coup for Brazil.

In contrast, there are only circa a total 7,000 ISPs in other Low- and Middle-Income Countries (LICs/LMICs) *combined*[474] – and yet Brazil *singularly* can boast of 20,000. This is why I strongly argue that democratising and localising connecting the next 2.6 billion through high-potential ISPs is exactly what is needed to connect more of the unconnected 2.6 billion unconnected to the Internet today.

## 10.4 Meaningful Connectivity Must be Taken Seriously by the ITU

I have already summarised 'meaningful connectivity' in Section 10.1, but I add that it must be taken seriously by the ITU – in particular. I ask the following question semi-rhetorically: does the ITU *genuinely* believe that the bulk of the 'Mid-tail' 2.7 billion (see Figure 10.1) are 'meaningfully' connected? My clear answer is No! The Alliance for Affordable Internet (A4AI) developed a well-

---

[474] Source: Paul Garnett, https://www.vernonburggroup.com/connect-one-billion

researched *Meaningful Connectivity framework*[475] and applied it to nine LMIC economies (Colombia, Ghana, India, Indonesia, Kenya, Mozambique, Nigeria, Rwanda, and South Africa). They used *real* mobile phone surveys to estimate the number of people with meaningful connectivity in each of these LMIC countries[476]. It is A4AI's findings and insights that I used in this book to estimate the number of people <u>not</u> meaningfully connected at more than 2 billion people globally (see Table 6.2).

I express my concern in this book that the ITU [and the Broadband Commission] are deliberately using a definition of being connected to the Internet [online] which significantly 'increases' the Mid-tail 2.7 billion 'connected' number in Figure 10.1 – even though most of these 2.7 billion are *not* 'meaningfully connected'. It feels cynical – and the ITU must *not* be accused of cynicism!

## 10.5 Two key Warnings to LDC Connectivity Leaders

In this book, I provide key warnings to Presidents, ICT Ministers, key ICT policy makers and regulators of LDCs, LMICs and LICs. They may not be that surprising given what you have read in this chapter so far – here they are:

a.  *Don't wait up for the Telecoms Market to Solve the 2.6 billion Connectivity Gap:* there are simply, increasingly no more market-led telecoms solutions to connect the tail 2.6 billion. This is to some extent echoing Webb's (2024a) 'End of Telecoms History' thesis.

    Recall that the telecoms market has 'failed' key [rural] populations – with a huge usage gap of 1.3 billion [minimum] *not* using basic [2G] voice services. 'Failed' – i.e. in quotes – because it is arguably unfair to put all or most of the blame at the foot of the telecoms industry, when usage gap issues like poverty, unaffordable phones, excise duties on phones, etc., are more in the control of governments. However, the fact is that the 1.3 billion have been 'failed' nonetheless.

---

[475] The framework focuses on four pillars: 4G-like speeds, smartphone ownership, daily use, and unlimited access at a regular location, like home, work, or a place of study.
[476] https://a4ai.org/report/advancing-meaningful-connectivity-towards-active-and-participatory-digital-societies/

b.  *Beware of hype around new technology solutions*: In this book, I advise LMIC/LIC/LDC leaders to be cautious of the promises made by 'sexy' tech solutions claiming to connect 2.6 billion unconnected people. The reader may, or would likely have, heard of technologies like 5G, perhaps even 6G, LEOs, MEOs, HAPs, HIBs, TVWS, optical fibre, or projects like Starlink, Oneweb, Project Kuipers, etc. The reality is that the digital divide continues to persist and arguably even gets worse with these solutions.

Rather, LDC leaders should seek to evolve more pragmatic, need-centred, and local solutions like M-Pesa in Kenya (see Section 5.1) – or Brazilian-type WISP solution providers, wherein 16.37% of Brazilian households in 2023 accessed broadband *by sharing a fixed broadband connection* with a neighbour (see Table 5.1). This is ingenious, local and pragmatic to the Brazilian context.

LDCs leaders own the connectivity digital divide in their countries just as much as they own patently unrealistic UN 17 Advocacy Targets[477], allegedly to be realised by 2030.

## 10.6 What Happens When Connectivity Markets Fall Short?

So, what should happen when markets fall short, and it is not the 'fault' of markets? I make the case all through the book that when markets and competition between its participants fail to produce positive outcomes for consumers and citizens – i.e., a significant market failure has ensued which has failed millions or billions of citizens – regulation (or regulatory intervention) becomes necessary to ensure net benefits. This is illustrated in Figure 6.4.

So, in this book, I overview India's BharatNet 'interventionist' connectivity programme – arguably the world's largest rural connectivity initiative. The programme was first set up in 2010 to bridge the digital divide across India when digital adoption in India was low, with total Internet users around 92 million. The programme has since contributed to a remarkable increase to 900 million

---

[477] What are the Sustainable Development Goals? | Policy and advocacy | Sightsavers

users by 2025. So, how did India achieve this? The short answer is via the following:

a. *Concerted policy leadership complemented by sustained regulatory interventions:* sustained digital divide policy over three phases to date since 2010 (Phases 1, 2 and 3) has been led and administered by the Department of Telecommunication (DoT)[478]. The BharatNet also benefited from significant regulatory interventions and funding from the Indian Universal Service Obligation Fund (USOF[479]). The USOF had been created to promote equitable access to telecommunication services across the country, particularly to bridge the digital divide in remote and rural areas. In this book, I emphasise the role of approaches and tools like techno-economic modelling (see Section 7.2) and Rural Connectivity/UAS Portals (see Figure 8.3), respectively. More of such regulatory interventions would sadly be needed to connect the 2.6 billion unconnected. As I note earlier, if market competition fails, regulatory interventions become more and more necessary (see Figure 6.4).

b. *Concerted implementation strategy using multiple models and technologies*: the BharatNet programme has been implemented employing multiple implementation models, including State-led, public sector/CPSU[480], private sector, and satellite connectivity approaches.

c. *Concerted and sustained leadership*: despite the many limitations of the BharatNet programme that I describe in Chapter 7, it is unarguable that the digital divide leadership has been both concerted and sustained. As with many aspects of society, there is no replacement for effective and inspired leadership.

So, reducing the voice/data digital divide and improving meaningful connectivity requires: (i) *Policy & Regulation* to propose clear policy and regulatory directions; (ii) *Implementing Strategy;* (iii) *Policy & Regulation;* and (iv) *Leadership.*

---

[478] "BharatNet: Bridging the Digital Divide." Accessed: May 10, 2025. [Online]. Available: https://pib.gov.in/PressReleseDetailm.aspx?PRID=2086701&reg=3&lang=1
[479] "USOF Ongoing Schemes - Advancing Telecom Connectivity in India." Accessed: Feb. 10, 2025. [Online]. Available: https://usof.gov.in/en/ongoing-schemes
[480] Central Public Sector Undertakings (India)

In this book, I overview the many categories of leadership that are all involved in some way in bridging the connectivity digital divide across the globe. This is important because the leadership of connecting the 2.6 billion across the globe gets quite confusing.

## 10.7 FIVE Proposed Initial Actionable Steps and by Who

*Sometimes the smallest step in the right direction ends up being the biggest step... Tiptoe if you must, but take a step*
— *Naeem Callaway*

In this entire book, I have humbly presented my "manifesto" for bridging the digital divide. So, as I draw to the conclusion of this book, what then are some clear action points and timescales? Can the key ones be tabulated, showing what is needed and who is responsible for doing it, and ideally why it will work this time round? Permit me to venture my initial top-5 actionable steps. I went for a minimalist number, i.e. just 5, such that they could be debated, updated and implemented in short order.

So, I present next a draft table that *transforms* some of the key insights and recommendations from this entire book into several initial and actionable steps with timescales, responsibilities, and rationales. It is a first draft action table tailored to reflect the global and national digital divide leadership next steps that I think are required to start tackling the nub of the Connectivity Crisis. I believe and recommend that the responsible parties I identify in Table 10.1 execute these actions forthwith. Tiptoe if they must, but they should take these initial steps as the Naeem Callaway quote I use to start this section posits.

Table 10.1 – Five Proposed Actionable Digital Divide Next Steps

| Proposed Action Point | 1) Establish 'fit-for-purpose' National Divide Leadership & Governance |
|---|---|
| Timescale | 6 – 12 months (from kick-off) |
| Responsible Parties (proposed leads and co-leads are <u>underlined</u>) | <u>National Governments (Digital Ministries & Telecom Regulators)</u> with Civil Society (NGOs) Input – and inputs and insights based on this book facilitated by Consultants[481]. |
| *Why This Will Work...* | |

- National Governments must FULLY <u>own</u> their digital divide challenges – NOT waiting for all 'to be okay on the night'. True digital divide, universal connectivity leadership across LDCs, LLDCs, LICS, LMICs and SIDS nations is very lacking. ALL LDCs (in particular) must take the problem seriously with <u>authentic</u> digital divide leadership (cf. Section 9.1).
- Country-level Digital Leadership Platforms would emerge and be piloted.
- Crafting a truly 'living' Country-led Mitigating Digital Divide Strategy would make a big difference (cf. Section 7.1).
- LDCs **would learn** from India's BharatNet (cf. Section 7.2).
- Indeed, **drawing from the insights and recommendations in this book** – if fully 'digested' at national level – would lead to much better national digital divide leadership, governance and execution.

| Proposed Action Point | 2) Establish a Global Digital Divide Leadership Coordination Forum (GDDLCF) |
|---|---|
| Timescale | 6 – 12 months (from kick-off) |
| Responsible Parties (proposed leads and co-leads are <u>underlined</u>) | <u>UN Broadband Commission</u> / <u>ITU</u> / LICs/LMICs Representatives |
| *Why This Will Work...* | |

- The most digitally unconnected get key representative 'seats at the table'.
- Ensuring that the realities of LIMCs/LICs digital divide root causes are fully appreciated, cf. Sections 2.2 & 2.3.
- Centralizing coordination reduces counterproductive competition between leadership tiers.
- GDDLCF would prevent siloed efforts and "leadership fragmentation" by setting a clear agenda early.
- GDDLCF **would provide** Annual Accountability Reports on Bridging the Divide. Regular reporting from multi-stakeholder Panels **would build** trust, measures progress, and identify gaps.

---

[481] Like Cenerva Ltd (www.cenerva.com) or other similarly-qualified digital divide consultancies.

| Proposed Action Point | 3) GDDLCF to frame anew "The Connectivity Crisis - Half the World Left Behind" problem statement with more realistic UN Connectivity deadlines and timescales |
|---|---|
| Timescale | 6 months after GDDLCF established |
| Responsible Parties (proposed leads and co-leads are underlined) | <u>UN Broadband Commission</u> / <u>ITU</u> / <u>GDDLCF</u> / LICs/LMICs Representatives / Regional Blocs / Major financial donors |

*Why This Will Work...*

- "A problem well stated is half solved."— Charles Kettering.
- Acknowledge 2.6 Billion *unconnected* and a further 2 billion *under connected* – as shown in Figure 10.1.
- **ITU would redefine what meaningful connectivity is** – certainly away from "in the last 3 months" (cf. Section 1.2).
- **ITU would acknowledge firmly the usage-gap/demand-side affordability**, digital literacy, and socio-economic factors[482], leading to even greater spending of billions of dollars on demand-side digital divide solutions, rather than just on supply-side.
- ITU and ITU Commission **would learn lessons** from Pareto (Section 1.3), After Access Surveys[483], A4AI's Meaningful Connectivity framework[484], BharatNet (cf. Section 7.2), etc. to reframe the connectivity challenge and realistic UN/ITU deadlines.

| Proposed Action Point | 4) GDDLCF to lead definition of Roles for Each Leadership Category: Establishing global connectivity leadership consortium with clearer and codified responsibilities covering *Policy, Collecting Evidence, M&E, Regulatory, Technology, Pro-connectivity Standards, Funding, Collaboration*, etc. |
|---|---|
| Timescale | 12 months after GDDLCF established |
| Responsible Parties (proposed leads and co-leads are underlined) | <u>UN Broadband Commission</u> / <u>ITU</u> / <u>GDDLCF</u> / LICs/LMICs Representatives & Digital Ministries / MDBs / Regional blocs / Other Major financial donors / Big Telco / Big Tech / GSMA / International Economic Forums / Regional Telecom Alliances / NGOs & <u>Local Leaders</u> / Academic Consortia |

---

[482] https://afrisig.org/sites/default/files/pdf/Acess-After-Access-AfriSIG-2024.pdf
[483] https://researchictafrica.net/project/after-access-2022-survey/
[484] The framework focuses on four pillars: 4G-like speeds, smartphone ownership, daily use, and unlimited access at a regular location, like home, work, or a place of study.

- Digital divide leadership nationally and across the globe gets quite confusing with much duplicative, overlapping and non-collaborative activities (cf. Chapter 9).
- Clarity on roles prevents national strategic manipulation and misalignment.
- Eliminate/minimise some LDCs 'playing' off one or more digital divide category/categories of 'leadership' players against others.
- Formal structures prevent "leadership gaming" and boosts cooperation.
- ITU continues to lead on *Policy, Collecting Evidence and M&E* – but including LDC Nations defining their own Digital Divide Policies (cf. Section 8.2); another grouping of Technology stakeholders on digital divide solutions with traits as described in Chapter 3, like Civilisation in a box (Chapter 4); another grouping led by [India] on a more pro-connectivity mobile standard (cf. Section 7.4); a Connectivity Regulatory Forum to implement as Chapter 8 recommends (cf. Sections 8.2 and 8.3); another grouping on a more joint-up Funding for digital divide activities including addressing the "fair share" challenge (cf. Section 9.1); another grouping on Collaborations to truly speed up digital divide projects worldwide and regionally; Etc.
- The GDDLCF **would** Launch Connectivity Investment Taskforce as per last bullet.
- **Regional Connectivity Pilot Projects would and should be developed and launched** to demonstrate scalable models and build local buy-in, e.g. Connectivity-in-a-Box.
- Such a **global connectivity leadership consortium** would provide the true coordinated leadership with the likes of the World Bank, Private Sector Coalitions, Donors, etc. – a joint financial strategy can align resources, unlock "fair share" contributions.

| Proposed Action Point | 5) Create a G7[485] / G20[486] Multilateral 'Connectivity Pact' Charter that would enable other key digital divide activities. *[Both the G7 and the G20 are fora for international cooperation that shapes the governance of major social and economic issues].* |
|---|---|
| Timescale | 12-24 months after GDDLCF established |

---

[485] The G7 is an intergovernmental forum made up of Canada, France, Germany, Italy, Japan, the UK, and the US. The European Union also participates but is not an official member.

[486] "The G20 comprises 19 countries (Argentina, Australia, Brazil, Canada, China, France, Germany, India, Indonesia, Italy, Japan, Republic of Korea, Mexico, Russia, Saudi Arabia, South Africa, Türkiye, the United Kingdom, and the United States), the European Union, and since 2023, the African Union." - Source: https://g20.org/about-g20/g20-members/.

| Responsible Parties (proposed leads and co-leads are underlined) | **G7/G20 Digital Alliances** / **ITU** / Regional blocs / GDDLCF / ITU Broadband Commission / LICs/LMICs Representatives |
|---|---|
| *Why This Will Work...* | |

- G7/G20 Digital Alliances – which are already evolving today – **would be critical in helping DRIVE the previous 4 actions.**
- The G20 plays a crucial role in bridging the digital divide by promoting *inclusive digital transformation*, ensuring equitable access to technology, and fostering global cooperation on digital infrastructure. "G20 members include the world's major economies, representing 85% of global GDP, 75% of international trade, and two-thirds of the world's population"[487].
- G7/G20 Digital Alliances **would encourage alignment** between State and non-State leaders – and build policy reciprocity.
- G7/G20 – as a more competent Multilateral and Intergovernmental leadership grouping – may/**would be in a better position to work with MDBs**, donor blocs, think thanks to design a Fair-Share Investment model – and bring together Big Tech and Big Telco.
- A Global Digital Divide Investment Needs Tracker would be developed – with biannual updates from the G7/G20 & ITU in order to "ground" the 2.6 billion unconnected challenge in financial realism and urgency.

These initial and minimal actions will serve as a starting point for reframing and resetting the global approach to addressing the Connectivity Crisis by key stakeholders. Other chapters in this book (specifically Chapters 7 to 9) provide numerous detailed recommendations.

As I note at the start of this section, I recommend that the responsible parties identified in Table 10.1 promptly execute these actions. *If necessary, they may tiptoe if they must*, as emphasised by the Naeem Callaway quote at the beginning of this section.

## 10.8 Final Summary

I note earlier in this chapter that just as this book was going into final publication, the ITU published their September 2025 'Connecting humanity action blueprint: advancing sustainable, affordable and innovative solutions' report[488]. The

---

[487] Source: https://g20.org/about-g20/g20-members/
[488] https://www.itu.int/dms_pub/itu-s/opb/gen/s-gen-invest.con-2025-pdf-e.pdf

conclusion of this seminal report includes: "Achieving universal, meaningful Internet connectivity by 2030 will take the mobilization of private sector companies, governments, civil society, international organizations and DFIs that are invested in closing the digital divide". Of course, I broadly agree with the latter as I cover in the previous section and in more detail in Chapter 9. However, I only broadly agree partly because the ITU is still maintaining 2030 is credible - even if it is 'aspirational'. My Pareto prediction of 86% effort levels still being ahead of us should put a pin in this 2030 date, and it is further incredulous to me the ITU aspires to meaningful internet connectivity by this date too. I have strongly argued in this chapter and entire book that the ITU urgently needs to review what it currently defines as 'meaningful internet connectivity'. In addition, even if this report's estimated USD $2.6 trillion to $2.8 trillion required to close the digital divide was magically conjured up, this 2030 date would still never be realised as my Malawi and PNG case studies in Chapter 2 'prove'. It is my genuine wish for the ITU to consider the messages and lessons from this humble book.

Overall, I hope the audience I intend for this book (who I list in Section 1.5 of Chapter 1) will seriously consider, 'adopt' and 'own' the insights, evidence, argumentations, guidance and recommendations of this book towards bridging the digital divide of the 2.6 billion unconnected, and meaningfully connecting circa 2 billion more. These goals are not only noble but are truly non-trivial to achieve. This entire book is a distillation of insights and evidence I have learnt from working with developing countries, and hailing from one myself.

It is therefore a personal and honest book. It means *no harm* to the many stakeholders trying their utmost to address the universal connectivity challenges.

I hope to have provided a range of [new] perspectives on the challenge of universal digital connectivity, drawing from Kettering's principle that "a problem well stated is half solved". My insights are informed by principles such as Pareto's law, metaphors from basic physics like pulleys and levers, and evidence from reputable sources such as the ITU, GSMA, A4AI, among others. I then derive and present distinct and hopefully unique insights, advice, and guidance and recommendations for developing countries, the Broadband Commission, the ITU, the World Bank Group, UN agencies, and more. Ultimately, the assessment of my success in achieving the ambitious objectives set forth in this book lies with you, the reader. Where I may fall short, I encourage you and others to build upon my work. This is the true essence of scholarship.

Clearly, articulating and reformulating complex social and economic problems is essential. It is possible that talented students, engineers, entrepreneurs, or established organisations, particularly in emerging markets such as the BRICS countries, may find inspiration in the insights presented in this chapter and throughout this book. This could motivate them to revolutionise the telecommunications industry to better serve currently unconnected and under connected populations, which would be a positive development.

It is my sincere hope that – '*The Connectivity Crisis - Half the World Left Behind*' – will one day be a matter of history. However, I acknowledge that this is not an imminent reality.

# About the Author

## H Sama Nwana[489], CITP, FBCS, FIET, CEng, BSc (Hons), MSc (Dist.), PhD, MA (Cambridge), MBA (Dist.) (London Business School)

H Sama Nwana is Managing Partner of Cenerva Ltd[490], a boutique UK-based training-led consultancy based in London (UK) on Telecoms, Media, Technology (TMT), Digital Economy and General Cross-Sector Network Industries regulatory issues, with an emphasis on developing and emerging markets. He is also a Full Visiting Professor at the University of Strathclyde (UK) and has held similar roles in the past at the Universities of Bristol (UK) and Brunel University (UK).

---

[489] (99+) H Sama Nwana | LinkedIn - https://www.linkedin.com/in/h-sama-nwana-bbb0742/?originalSubdomain=uk
[490] www.cenerva.com

Nwana has published other authoritative books relevant to this one, including:

- *Demystifying Economic Regulation: A Practitioner's Guide: Theory, Methods and Practice* – published in August 2024.

- *The Internet Value Chain & the Digital Economy: Insight and Guidance on Digital Economy Policy and Regulation* – published in April 2022; and a follow-on book

- *Telecommunications, Media & Technology (TMT) for Emerging Economies: How to make TMT Improve Developing Economies for the 2020s* – published in April 2014

In fact, he is the author/editor of 9 books (including this one) and more than 120 refereed papers in journals, conferences and industry publications. He is a heavily cited author according to Google Scholar with an i10-index of 51 and more than 9000 citations – considered very high for a non-academic author[491].

As a senior executive board member, former senior regulator at Ofcom UK, ex-industry MD, multiple award-winning technologist, and thought leader, he regularly consults[492] and delivers C-level trainings for the likes of IFC, World Bank, Meta/Facebook, Microsoft, MTN, USAID, Australia Aid, as well as Governments, network regulators, competition authorities, venture capital firms and telecoms operators across the Caribbean, Middle East, Europe, South East Asia and Africa. Hailing from Cameroon, and hence being African, he is passionate about connecting Africa's/ASEAN's billions of unconnected through a combination of entrepreneurial, commercial, regulatory and policy instruments – as this book attests. In February 2024, he joined the Board of the Global Cyber Alliance (GCA)[493]. He also sits on the Advisory Committee of the IEEE[494] Connecting the Unconnected, an IEEE Future Networks Program[495].

---

[491] H S NWANA - Google Scholar -
https://scholar.google.fi/citations?user=UIiDJjAAAAAJ&hl=th
[492] Or has done in the past.
[493] https://globalcyberalliance.org/
[494494] The Institute of Electrical and Electronics Engineers is an American 501 public charity for electrical engineering, electronics engineering, and other related disciplines.
[495] https://ctu.ieee.org/our-team/2023-team/advisory-committee-members/

# Selected References & Bibliography

A4AI (2022), *Advancing Meaningful Connectivity: Towards Active & Participatory Digital Societies*. Alliance for Affordable Internet.

Euler, S., Lin, X., Tejedor, E., & Obregon, E. (2021). *A Primer on HIBS – High Altitude Platform Stations as IMT Base Stations*. arXiv. https://arxiv.org/abs/2101.03072

Nahm, W, (2023) "Satellite internet technology: A double-edged sword", *Research Outreach*, **138.** DOI: 10.32907/RO-138-5356635156

Nwana, H. S. (2014), *Telecommunications, Media & Technology (TMT) for Developing Economies: How TMT can Improve Developing Economies in Asia and Elsewhere for the 2020s*, London: Gigalen Press, http://www.amazon.co.uk/Telecommunications-Media-Technology-Developing-Economies/dp/099282110X.

Nwana, H. S. (2018), *HAPS for Affordable Broadband Connectivity*, May 2018, from https://cenerva.com/resources/white-papers/

Nwana, H. S. (2022), *The Internet Value Chain and The Digital Economy: Insight and Guidance on Digital Economy Policy and Regulation*, London: Pita Press, The Internet Value Chain and The Digital Economy: Insight and Guidance on Digital Economy Policy and Regulation (Telecoms, Media & Technology - Digital Economy): Amazon.co.uk: Nwana, H Sama: 9798800751444: Books

Nwana, H. S. (2024), *Demystifying Regulation*: *A Practitioner's Guide: Theory, Methods and Practice*, Glasgow: Strathclyde Academic Media.

Popescu, A., Erman, D., Ilie, D., Fiedler, M., Popescu, A., de Vogeleer, K. (2011), "Seamless Roaming: Developments and Challenges", in: Kouvatsos, D.D. (eds) *Network Performance Engineering. Lecture Notes in Computer Science*, vol **5233**. Springer, Berlin, Heidelberg. https://doi.org/10.1007/978-3-642-02742-0_34

Resnick, D. (2020). *The politics of resentment: Opposition mobilization in Africa's electoral autocracies*, *African Affairs,* **119**(476), 27–48, Oxford Academic Press.

Toure, Hamadoun (2013), Investing in Broadband Infrastructure, in *Africa Infrastructure Investment Report 2013*, Commonwealth Business Council, March 2013, pp. 169. https://commonwealthbc.com/africa-infrastructure-investment-report-2013/

Webb, W. (2018), *The 5G Myth: When Vision Decouples from Reality*, Germany: De G Press, 3rd Edition.

Webb, W. (2024), *The 6G Manifesto*, ISBN - 979-8338481

Webb, W. (2024a), *The End of Telecoms History,* ISBN-13: 979-8328402729, https://www.amazon.co.uk/dp/B0D83Z5FYJ/

# Index